BIRD TALK

BIRD TALK

AN EXPLORATION OF AVIAN COMMUNICATION

Barbara Ballentine and Jeremy Hyman

Consultant Editor
Mike Webster

CSIRO

PUBLISHING

A catalogue record for this book is available from the National Library of Australia.

ISBN: 9781486315307 (hbk)

This edition published exclusively in print only,
in Australia and New Zealand, by:
CSIRO Publishing
Locked Bag 10
Clayton South VIC 3169
Australia
Telephone: +61 3 9545 8400
Email: publishing.sales@csiro.au
Website: www.publish.csiro.au

This book was designed and produced by
The Bright Press, an imprint of The Quarto Group
The Old Brewery, 6 Blundell Street
London N7 9BH, United Kingdom
T (0)20 7700 6700
www.QuartoKnows.com

PUBLISHER James Evans
EDITORIAL DIRECTOR Isheeta Mustafi
ART DIRECTOR James Lawrence
MANAGING EDITOR Jacqui Sayers
COMMISSIONING EDITOR Kate Shanahan
PROJECT EDITORS Joanna Bentley & David Price-Goodfellow
DESIGNER Wayne Blades
PICTURE RESEARCHER Alison Stevens

Printed in Singapore

10 9 8 7 6 5 4 3 2 1

Cover photos Front: Andrew Parkinson/naturepl.com,
back: Nicolae Cirmu/shutterstock.

CONTENTS

FOREWORD

Spending most of my days at home this past spring, I took a stroll around the property every morning, noting all of the species of bird that I could see or hear. As March wore on into April, and then on into May, those walks became filled with more birds, more action and more sound. A sapsucker arrhythmically drumming on a tree (a sure sign that spring is coming here in central New York state). The sounds of air rushing through the feathers of a woodcock in flight. The stunning, retina-burning colours of the warblers freshly back from their long migration. And the honking of geese flying high overhead as they headed northwards. The woods around my home were throbbing with the sights and sounds of dozens of birds.

That is the thing about birds: they're just so obvious. In this respect birds differ from most other animal groups. For example, most mammals (think of a mouse or bat) spend the vast majority of their time trying not to be noticed – they hide, and slink, and stay quiet in dark corners. But birds don't do that. Birds sing and call from treetops. They flash brightly coloured plumage. They jump and dance and cavort with elaborate displays. Birds stand out. And they do so because they are constantly 'talking' to each other. Conversing with their bright colours, with their amazing voices and even in some cases with their odours. Birds are constantly communicating.

But what are they saying, and why are they saying it? Are they born knowing their own language, or do they learn how to call to others in their flock? And what intricate mechanisms are responsible for the amazing colours, displays and songs that birds produce? Questions like these have been a central focus of research in animal behaviour for decades, and the scientists doing that work, using both sophisticated and sometimes surprisingly simple experiments, have uncovered a lot about how and why birds talk to each other.

This book is about that research. It is about what we have learned regarding the ways that birds communicate with each other; the ways that a bird talks to its mate, to its offspring, to its flock mates and even to the predators that want to eat it. This book is also about how we have learned all of this: the science behind the knowledge. Richly illustrated and engaging, this book is an excellent introduction to the fascinating world of the birds around us.

Spring and summer have now passed, and I am still kept company by the birds. The migrants have headed back south, but the woods are still alive. The chickadees still call from the trees (mostly now the tell-tale 'chick-a-dee-dee-dee' call rather than the two-note territorial song), the owls still call soothingly in the darkness, and the geese still honk high overhead, though now heading southwards. The birds are still talking to each other despite the changes in the world around us. I find that very comforting.

Mike Webster
Ithaca, New York

The Yellow Warbler is a common and widespread visitor during summer in North America.

INTRODUCTION:
WHAT IS COMMUNICATION?

When birds sing, call and show off flashes of colour, they are using signals and engaging in communication. Communication is essential for birds to navigate the social interactions that are necessary for their success. Signals allow birds to communicate complex information with one another about things such as the presence of a threat, their health, status, motivation and their identity to other birds.

INFORMATION

Signals have been shaped by evolution to contain information, and communication is the process by which multiple individuals exchange that information using signals. Every communication requires a signaller – the individual that produces the signal – and a receiver – the individual that responds to the signal. Humans are familiar with communication using a variety of signals and in different contexts, and birds communicate in ways and about subjects that should seem familiar to us.

In some situations, signallers and receivers have common interests, such as protecting relatives from danger. A signal might provide information about predator threats in the area, for instance. The benefit to the signaller is that their mate, offspring or related flock mates are alerted to danger and their likelihood of survival increases. The benefit to the receiver is essentially the same. In such a situation, we expect most communication to be honest: natural selection favours signallers that give good, reliable information about the presence of predators, and natural selection favours receivers that respond to reliable signals.

When signallers and receivers are in competition, such as two territorial birds fighting over resources, signals might provide information about fighting ability or the likelihood of an attack. The benefit to the signaller

⊙

Whether they are flirting or fighting, these two Hawfinches are engaged in communication, and are probably exchanging vocalisations as well as showing off aspects of their plumage.

is clear: if it can convince a rival of its superior fighting ability, then it can gain access to valuable resources without resorting to fighting. But why should receivers pay any attention to the signal? A receiver that retreats from every threat will not get the resources it needs to survive. A receiver that disregards every threat will get into fights it is unlikely to win.

When signallers and receivers have different interests, such as males advertising themselves to females while females act as choosy consumers, signals might provide information about the health, vigour or parental ability of the males. But males are often signalling to attract as many females as possible and females are attending to signals in the hope of mating with the highest-quality male. Again, the benefit to the signaller is clear: if he can convince a female, or multiple females, that he is the highest-quality male, then he will produce numerous offspring, and send more copies of his genes to the next generation. But why should receivers pay any attention to the signal? If females are unable to directly and quickly assess male quality, a signal that is related to male quality can give her important information that improves the quality or quantity of her offspring. Attending to male signals can be costly for females, however – they may spend considerable time and energy surveying advertising males. And, if a female is too choosy, she may not mate at all.

Even when signallers and receivers are in conflict, while signallers could benefit from deception, receivers will benefit from acquiring reliable information. As a result, we would expect most signals to be honest. A widespread theory in evolutionary biology is that, when signallers and receivers are in conflict, signals are honest when they are also costly.

Signals can be costly for a variety of reasons. Morphological traits, such as colourful plumage, or behavioural traits, such as songs and dances, are often costly to produce or maintain – for instance, producing large, colourful feathers requires a lot of energy and nutrients. Owing to these costs, signal expression may tend to correlate with the quality of the signaller.

And, by paying attention to the signal, the receiver is likely to gather correct information about the health or vigour of the sender – less healthy or vigorous individuals cannot afford to produce the most extreme version of the signal. If this were not true, then birds wouldn't really be communicating – all the whistles, trills and chips would not be signals, they would just be noise.

To fully understand signalling in birds it is necessary to understand: (1) the underlying theory that predicts how signals could evolve to be optimally functional in the environment, (2) how senders use signals to transmit information to an intended audience, (3) how receivers perceive the information content of a signal and (4) whether we would expect signals to evolve to provide reliable information to receivers about senders.

⊙ Blue-footed Boobies use their bright blue feet as signals in communication.

⊙ Vocalisations and postural displays can be combined, as in this threesome of Magellanic Oystercatchers.

⊙ The delicate plumes of a Great Egret are well-known for their use as a signal in courtship, and the green facial skin of this bird is probably a signal of reproductive condition as well.

SIGNAL DETECTION THEORY

Understanding the evolution of the signalling systems in birds and other animals began with the development of signal detection theory (SDT). This theory arose from attempts to determine signal from noise in radar transmissions of enemy aircraft during World War II. The basic idea of SDT is that signals should stand out sufficiently from background noise to maximise correct detection while minimising false alarms. The ability to detect a signal can be impacted by extraneous information – the noise – that masks the signal, decreasing the probability that the signal would be detected correctly. Researchers adopted SDT in the 1950s to understand sensory psychology and, later, sensory ecology in animals.

SDT provides a framework for understanding how signals evolve in animals to maximise detection, minimise false alarms and ultimately provide reliable information to receivers. SDT helps us to understand three key issues in communication: (1) how environmental noise influences the production of signals, (2) how noise might influence the investment of signallers in their signals and (3) how noise might influence the responsiveness of receivers to signals.

Imagine the common scenario of trying to have a conversation with a dinner partner in a restaurant filled with other diners. Hopefully, you and your dinner partner trade off the roles of signaller and receiver. The accumulation of other conversations, music and noise can prevent you from communicating properly with your dinner partner. As a signaller, you are less able to communicate unless you exaggerate your signal, by shouting or repeating yourself, but that exaggeration can come at a cost – shouting can be exhausting. As a receiver, you must make a greater effort to detect or understand a signal masked by noise. As a signaller, if you have nothing interesting to say, you are less likely to invest in shouting. As a receiver, if your partner has nothing valuable to say, you are less likely to invest in listening. However, inattention can come at a cost, because you might end up missing the one valuable thing that your dinner partner had to say all night. In the presence of noise, a receiver is inevitably going to make a few mistakes.

This scenario represents a problem faced by all animals trying to communicate. All communication takes place in noise – sometimes the noise of other signallers of your own species, and sometimes the noise of the environment, including other species and wind, for instance.

On an open perch, a Scarlet Macaw and a Great Green Macaw show off their bright colours.

If birds are investing in producing expensive signals such as vocalisations, colourful plumage or behavioural displays, then we would expect those signals to be detectable in the presence of environmental noise. Further, if a bird (the sender) is investing in an expensive signal, the signal should be of interest to other individuals (the receivers). Likewise, if a receiver is going to invest in attending to signals, there should be a benefit to the receiver, usually in the form of information that is beneficial for the receiver to know.

So, SDT predicts that natural selection should favour senders who produce signals that are easily detectable by receivers and contain information that is beneficial to the receivers. Because there can be significant costs and benefits for both the sender and the receiver to cooperate in a reliable signalling system, selection should also favour signallers that minimise false alarms and missed detections. So, when the benefits of attending to a signal – and receiving correct information – are high, natural selection should favour the production of signals that maximise receivers' ability to gather that information.

Most of the signalling discussed in this book – between parents and offspring, males and females, and competing males or females – have evolved to provide the receivers with reliable information about senders or situations. However, deception and manipulation can evolve, too, but only when the costs of producing a signal that will give false information are low, and the costs of being fooled by the signal are low. A good example of using acoustic alarm calls for deception and manipulation can be found in our description of how Fork-tailed Drongos use mimicry to fool competitors into abandoning a food item by persuading them to think there is a predator in the area (see page 142).

SDT also helps us to understand how signals are shaped by the environment to increase detection. The colours of bird plumage can serve multiple functions, from camouflage to sexual display. Camouflaged feathers, obviously, are meant to blend into the background. But feathers used in displays tend to include colours that contrast with the environment so that they stand out from the background. Similarly, acoustic signals will have properties that allow them to stand out from background noise. In addition, when an acoustic signal is produced, the habitat can degrade the signal in ways that make it more difficult to understand. For example, in closed environments, such as forests with a lot of reflective surfaces, birds' songs include notes with slower repetition rates so as not to be impacted by reverberation.

Reverberation causes the blurring of notes in a song and can reduce the probability that receivers will correctly detect any information from the signals. Again, think of a restaurant with a lot of hard, smooth surfaces – not only is the amplitude of the crowd noise higher, but as the noise bounces off the surfaces it produces reverberation, which makes it more difficult to hear words or listen to music. In fact, rooms built for listening to music performances – such as opera or symphony halls – or recording them are carefully engineered to reduce reverberation with sound-absorbing surfaces.

This Fork-tailed Drongo is hovering over a Meerkat and can produce 'false alarm' calls to scare the Meerkat away from a morsel of food.

 Though a Yellow-shouldered Amazon parrot may seem brightly coloured, among the green leaves of a tree canopy it is well camouflaged.

These Red-and-green Macaws stand out brilliantly against a green background.

SENSORY SYSTEMS

Visual and acoustic displays produced by senders depend on the sensory abilities of receivers. Thus, bird displays are tuned to the channels and at the sensitives of the receiver sensory systems. In this section, we discuss the two main sensory systems of birds that enable them to receive and decode visual and acoustic signals.

VISION

Bird vision, like all vertebrate vision, is sensitive to a very narrow range of cosmic electromagnetic radiation. The electromagnetic radiation receptors responsible for vision in vertebrates are found in the retina of the eye. Birds have two overlapping visual systems – scotopic and photopic. The scotopic visual system allows for visual acuity in dim light and this is provided by retinal cells called rods. The photopic visual system allows for the perception of colour by retinal cells called cones. Cones contain opsin, light-absorbing pigments, which allow cones to be sensitive to different wavelengths of light, or colours. Some cones contain additional oil droplets that further narrow the colour sensitivity.

Birds have tetrachromatic vision, meaning they have four different cone types that are sensitive to four different wavelengths of light: 370nm (violet), 445nm (blue), 508nm (green) and 560nm (orange). However, whereas human visual acuity includes wavelengths between 400nm and 760nm – we can see colours in the spectrum from violet to red – birds' visual acuity extends below 400nm to 300nm, which means they can see in the ultraviolet range as well.

The nerve impulses from the stimulated cells of the retina are carried to the brain and decoded into millions of different colours. Ninety-seven per cent of birds are diurnal and therefore have excellent colour vision, while nocturnal birds, such as the Brown Kiwi, have exchanged the ability to see colour with the ability to see well in dim light by having more rods and fewer cones in their retina.

Human and avian visual acuity compared
On the retina of the eye, humans have three cones that correspond to three peak wavelengths of colour perception: blue, red and yellow. Birds have four cones that allow for the extension of their visual acuity into the ultraviolet.

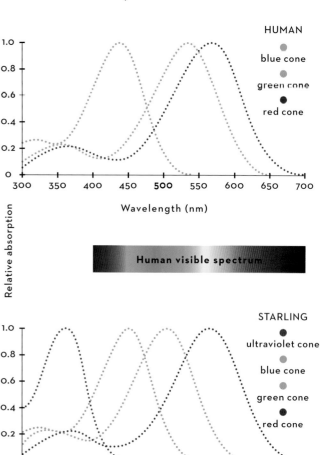

HEARING

Hearing physiology of the avian ear is simpler than the structure of a mammalian ear. Birds have one middle ear bone (the stapes) as opposed to three (the stapes, incus and malleus) found in mammals. Birds have a similar but simpler inner ear – the cochlea. The cochlea has hair cells surrounded by fluid that are sensitive to movement.

Sound pressure levels cause the tympanic membrane (ear drum) to vibrate, which transmits the vibrations to the middle ear bones, which vibrate against the cochlea at certain frequencies. The fluid in the cochlea transmits the vibrations to hair cells which are sensitive to certain frequencies. The hair cells send impulses to the brain which are decoded as sound frequency.

Despite having a simpler structure than humans', hearing acuity in birds and humans is similar. Like humans, most birds have the highest sensitivity at frequencies between 1–4 kilohertz (kHz). However, the range of sensitivity in most birds is a bit narrower. In general, we can hear what most birds are hearing. There are some notable exceptions. Owl species that hunt at night, such as the Barn Owl, have a higher sensitivity than humans and other birds across most of the range of frequencies, meaning they can detect sounds at very low amplitudes. Much of the difference in hearing sensitivities in owls is due to an extraordinary density of neurons in the brain devoted to processing acoustic information.

COMPARING HUMAN AND AVIAN HEARING ACUITY

Humans and birds are capable of hearing a similar range of frequencies. But, because humans are mammals, we have two more inner ear bones than birds and thus our hearing tends to be more sensitive. In the comparison shown below between humans and Zebra Finches, you can see that humans and birds have a similar frequency range of hearing ability. However, humans' ability to hear low frequencies is better, and across other frequencies humans can detect sounds at lower amplitudes (sounds need to be louder for birds to hear them). The peak sensitivity for both is around 4kHz. In bird species that have been tested, there is some variation in acuity among species, but the general trend in range of frequencies and peak sensitivity is similar.

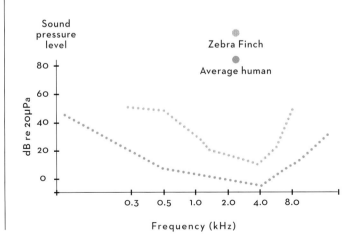

⬆

The Zebra Finch has become a model organism for laboratory studies of bird behaviour, genetics, anatomy and physiology. It is native to Australia, where it can be found in loud, active groups.

⬅

Human and avian hearing compared
Humans and birds are most sensitive to frequencies that range from approximately 2 to 4kHz. But, humans will hear those frequencies at lower amplitudes than birds.

1

COMMUNICATION CHANNELS

Communication in birds relies heavily on their sensory systems. Thus, communication channels are tuned to the anatomy and physiology of avian sensory systems. Birds have excellent vision and hearing and thus communication relies heavily on visual and acoustic displays. However, new research has found intriguing evidence that avian communication extends to olfaction as well. Understanding how birds can perceive and produce signals provides important insight into avian communication.

↑

The Grey Catbird sings
frequently throughout the
breeding season, but it is named
for its cat-like 'mewing' calls.

VOCALISATIONS

Birds produce a bewildering array of vocalisations from ethereal to harsh; guttural to melodic; simple to complex; sweeping tones to staccato trills. Thus, it is difficult to categorise bird vocalisations beyond the broad categories of songs and calls. Song is one of the more beautiful vocalisations produced by any animal, thus drawing our attention, admiration and imitation. In general, song is distinguished from calls because songs are often louder, longer and more complex than calls.

VISUALISING SOUND

Bird vocalisations research was revolutionised in 1954 by W H Thorpe who adopted the use of Bell Lab's sound spectrograph for detecting enemy submarines to studying bird song. The sound spectrograph allowed for the visualisation and quantification of recorded bird vocalisations, called 'spectrograms' or 'sonograms'. Spectrograms are generated from recordings and can be read like musical notation. The frequency, or 'pitch', of the sound is shown on the vertical or Y-axis, usually in units of kilohertz (kHz), and the length of the sound is shown on the horizontal axis, usually in units of seconds (s). Shading indicates amplitude (volume) in this view, the darker the area of the spectrogram the louder the sound (see box).

Researchers use spectrograms to determine how many song types or elements a male has in his repertoire, or to take measurements of acoustic features such as the highest frequency produced or the range of frequencies covered in a song, for example. Acoustic analyses of recordings of bird vocalisations have led to many discoveries about the development, production and functions of birds' songs and calls.

VISUALISING SOUND

Bird song (or any sound), once recorded and digitised, can be viewed in a variety of ways. All three views in this figure are of the same recording of a Song Sparrow song. The top view shows the relative changes in amplitude (loudness) across time. The trace in the middle view (a spectrogram) shows how frequency changes over time. When viewed together, you can see the frequencies of the sounds being produced and how loud they are relative to each other. The bottom view shows how much relative energy is present across frequencies of the song (i.e. which frequencies are most emphasised). In this case, the highest energy is at around 2kHz and likely represents the low-frequency whistles present at the beginning of the song.

Song Sparrow

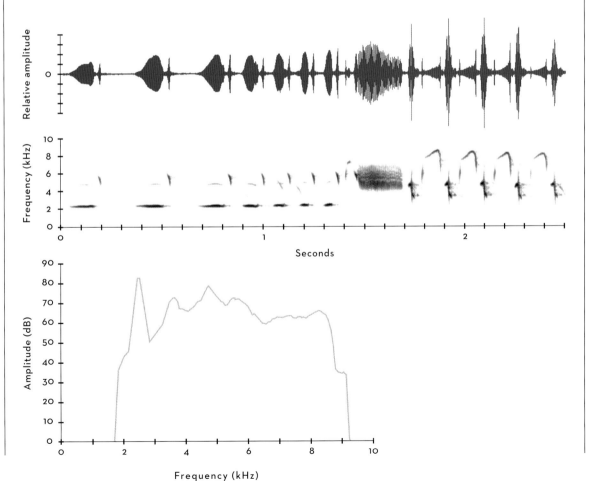

SOUND PRODUCTION

Birds produce vocalisations that typically range in frequency from several hundred hertz up to between 10 and 12 kilohertz. The range of frequencies of sounds produced by birds corresponds to the range of frequencies of sounds that can be perceived by birds. Acuity of hearing in humans and birds is similar so humans can hear most of the sounds produced by birds – although, some of those higher frequencies get harder to hear as we age.

The Brown Thrasher is a versatile singer, with a repertoire of more than 1000 song types. Ironically, this extreme vocal diversity has made it an unpopular subject for studies of bird song.

Veeries belong to a group of thrushes that can produce two notes simultaneously, one from each side of their syrinx, resulting in songs that have been described as flute-like, ethereal or haunting.

The way birds produce sounds, however, is very different to the way we produce them. Birds possess the larynx which we use to produce sounds, but birds use an organ called the syrinx for this purpose instead. In most birds the syrinx is located at the junction of the trachea and bronchi. The syrinx has a thin membrane on each side which is stimulated and controlled by muscles and nerves to vibrate at certain frequencies as air passes from the bronchi to the trachea. Sounds are produced by the frequency at which the membrane vibrates – the faster the vibration, the higher the frequency of sound. All birds have a syrinx, but some birds – for instance, vultures and ostriches – lack the necessary muscles to control vibrations of the membrane, and thus can only produce harsh sounds such as hisses and grunts.

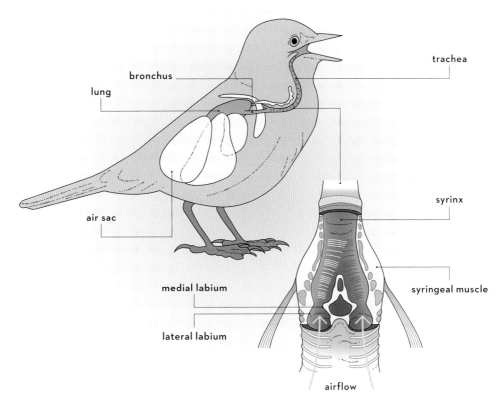

bronchus

lung

trachea

air sac

syrinx

medial labium

syringeal muscle

lateral labium

airflow

Vocal anatomy
The syrinx, specialised anatomy
for sound production found only
in birds, is named after a character
in Greek mythology who fashioned
reeds into panpipes.

Some groups of birds such as the oscines – the largest group of birds, with over 4000 species including many groups of small to medium-sized perching birds such as sparrows, thrushes and crows – have complex musculature that can independently control each side of the syrinx, enabling the production of complex sounds like song.

Because the syrinx has two membranes with independent control, birds essentially have two sources of sound production. In studies of syrinx lateralisation (contributions of each side), researchers have discovered that there are multiple (and complex) ways that the syrinx controls sound output. Bird anatomy can divide the labour of sound production between two sides of the syrinx. In many species one side of the syrinx will be dominant. Waterslager canaries primarily use the left side to produce song, while species such as the Zebra Finch exhibit right-side dominance. But, even when one side is dominant, the other may contribute. The studies on lateralisation of the syrinx revealed that the left side of the syrinx is limited to lower frequency sounds than the right side (hence the low-frequency songs of the

Waterslager canaries). Species such as Grey Catbirds and Brown Thrashers can alternate between different sides of the syrinx to produce alternating syllables or sounds. In broad range frequency sweeps, Northern Cardinals will use independent contributions of each side of the syrinx to produce a single note (left side for the low-frequency part, right side for the higher frequency part of the note). Species such as Wood Thrushes and Veerys can use independent control of each side of the syrinx to sustain two independent sounds simultaneously (two-voice syllables). Whichever arrangement of syringeal muscles a bird has, its sound production must be coordinated with breathing and often other parts of its anatomy, such as the bill.

SONG

The creation of sound, and especially song, involves the coordination of movements of the syrinx with control of breathing, and movements of the respiratory tract and beak. By elongating or shortening the trachea and by opening and closing their bills, birds can focus most of the energy of the sound at a single frequency to produce loud, pure-tone songs. Elongated vocal tracts favour low-frequency sounds and shortened vocal tracts favour high-frequency sounds. Vocal tracts in birds are analogous to instruments such as the trombone: when the slide is extended the instrument produces lower frequency sounds than when the slide is retracted. Some birds, such as cranes, have elongated tracheae that allow them to focus energy at very low frequencies (several hundred hertz). During song production, birds often move their bills in concert with the frequency of sound. The bill is most open when the highest-frequency sound is produced, and most closed when the lowest-frequency sound is produced. Coordinating all the moving parts of the body necessary to produce complex bird song is difficult and requires precise coordination and, in some cases, learning and practice.

This coordination of the vocal anatomy involved during song production is orchestrated by the brain. Birds have relatively large forebrains (as do humans), where much of the control and learning of song originates. The areas of the brain that are responsible for song production are influenced by hormones and can grow in the presence of testosterone and shrink in the absence of testosterone, so birds can grow new neurons at the beginning of every breeding season. In bird species that learn their songs, the size of the parts of the brain responsible for song learning and production correlates to the complexity of song – the higher the density of neurons, the more complex their songs.

Songs typically consist of a series of phrases or notes. Researchers refer to repertoires as either different renditions of songs or the notes and phrases used to make up a song. Types of notes that make up a song can be described as whistles (pure tone notes produced at nearly the same frequency), frequency sweeps (pure tone notes that cover a broad frequency range), trills (the same note or phrase repeated multiple times) and buzzes (broadband notes which sound harsh). Depending on the species, birds can assemble a song from a variety of sounds.

Pairs of Red-crowned Cranes produce vocal duets – a temporally coordinated series of male and female vocalisations.

The order of the sounds in a song can remain fixed or the elements of sound can be rearranged by the bird to make new versions of the song with each rendition. Song repertoires can be delivered by singing the same song type repeatedly and then switching, which creates eventual variety, or song repertoires can be delivered by switching between song types more often, producing immediate variety. Although song is often thought of as characteristic of male birds, females also sing in many species. In North America and Eurasia, there are many bird species in which females as well as well as males sing, but the vast majority of total song output that one hears comes from males. Species in which only males sing mostly exist in the Northern Hemisphere, but the situation is quite different in the tropics, where female solo song and precisely coordinated male and female duetting are common. Unfortunately, more bird song research has been done on Northern Hemisphere birds, so our understanding of female song lags behind.

↓

Neighbouring male Anna's Hummingbirds tend to sing more similar songs than non-neighbours – a pattern which is common in birds that learn their songs.

→

Blue-and-yellow Macaws are excellent vocal learners, but in the wild their skills are used to navigate their social relationships, rather than to entertain humans.

ROLE OF LEARNING

Song and call production in most groups of birds are thought to be mostly inherited rather than influenced by learning. However, there are three groups of birds for which we know the most about the influence of learning on certain types of vocalisations: song in oscines, song in hummingbirds and calls in parrots. It is thought that vocalisations are not learned in most other groups of birds. However, most of the research to date has focused on the development, production and control of song in oscines, known as songbirds.

The biggest group, consisting of over half of all bird species, are the passerines, which are divided into two groups – the oscines and the sub-oscines. The oscines are the true virtuosos of vocalisations because they produce some of the most complex and beautiful sounds in the animal kingdom. The oscines learn their songs, while sub-oscines, on the other hand, are traditionally thought to produce innate songs. However, research in multiple species of bellbirds (suboscines in the genus *Procnias*) suggests that male bellbirds are learning their songs. Research on learning in oscines has provided a framework for song learning in songbirds that is similar to learning language in humans. The stages of song learning are similar in all songbirds that have been tested.

But, the length of the stages can be variable across species. The entire learning process usually reflects the period that juvenile birds take to reach adulthood. In some species, the juvenile period is less than 1 year. But, in most species the juvenile period is about 1 year.

THERE ARE FOUR STAGES OF SONG LEARNING IN SONGBIRDS:

1. CRITICAL PERIOD Just like baby humans, most young birds like these Indigo Buntings must hear the songs produced by adults during a sensitive period. If they do not hear adult song during this period, they will be unable to produce normal songs as adults.

2. SILENT PERIOD This juvenile male Indigo Bunting during his first autumn would be starting to moult into breeding plumage but would not produce song during this period. The silent period is important for storing songs into memory for later comparison to practised song.

3. SUBSONG PERIOD At some point in the next weeks or months, depending on the species, birds will begin to practise the songs that are stored in their memories. Songs produced during this time lack the structure and timing of adult songs and the vocalisations are referred to as subsong. Males attempt to match the songs they produce with those stored in their memories. If birds are prevented from practising during this period they will be unable to produce normal songs. This stage is analogous to the importance of babbling in human babies as they begin to develop language. In their first breeding season, male Indigo Buntings have a mottled subadult plumage and still use subsong when they arrive at the breeding grounds for the first time. Indigo Buntings have a song crystallisation period that extends into their first breeding season. The extended period of song crystallisation allows Indigo Buntings to crystallise songs that match songs of neighbouring territorial males.

4. SONG CRYSTALLISATION Over time, practising singing results in the young birds producing adult-like songs with notes organised into the correct order, as well as timing that imitates the songs heard during the critical period. This Indigo Bunting is in full song given from an exposed perch. Not all the practised songs will end up in the repertoire of crystallised song. Birds can also improvise components of songs to produce brand new arrangements of songs or syllables that are different from the training songs. In some songbird species, in what is known as closed-ended learners such as Song Sparrow and Zebra Finches, once a song is learned, that is the song they sing for life. While in others – open-ended learners, such as Indigo Buntings – they can increase their vocal repertoire once they become adults. For some species that are classified as open-ended learners, it is unknown whether they are increasing their vocal repertoire by including songs memorised during the critical period, or if they are memorising new songs. We do know that in between the subsong period and song crystallisation, birds can cull song types from their subsong repertoire. It is possible that the open-ended learners increase repertoire size by including more songs memorised during the critical period. However, an example of a true open-ended learner is the European Starling, which can memorise and produce new songs at any point in its life. Errors in song learning lead to the formation of local song dialects, which we discuss later on (see pages 31, 57 and 166–7). Male Indigo Buntings returning for the second breeding season will have full adult plumage and have the fully crystallised song obtained during their first breeding season.

GENETICS AND SONG DEVELOPMENT

Although learning is important for song production in songbirds, there is also a genetic component to song production. Birds seem to have an innate template for development of species-specific song. Early research on song learning showed that, during the critical period, birds have a predisposition to learn the songs of their own species, suggesting song development is at some level genetically controlled. More recent studies have identified specific genes that are important in the formation of auditory pathways and contribute to species-specific song recognition and production. There is much left to learn, however, about the genetic control of song development and production in songbirds.

VOCAL MIMICRY

Most songbirds have an innate bias towards memorising songs of their own species. However, about 20% of songbird species copy the songs of other bird species, other animal species and even anthropogenic noise, including human voices. Some species like the Northern Mockingbird are well-known for imitation, but they are not especially good at it. The mimicked songs are produced at a distinct tempo that is characteristic of Northern Mockingbirds and not the species they mimic. Most birdwatchers would not be fooled by mockingbird imitations of local birds. But other species that use mimicry, like the Superb Lyrebird of Australia, are excellent mimics and can even produce accurate imitations of chainsaws and camera shutters, as well as the vocalisations of other birds and mammals. In the case of the European Marsh Warbler, a migratory species, the adults stop singing before the critical period begins in young birds. Therefore, European Marsh Warblers sing only the songs of other species with a repertoire that on average represents seventy-seven different species they hear during migration, including songs that cannot be assigned to any species. Parrots, of course, are renowned for their ability to imitate human voices. Other species can imitate human voices, but parrot imitations can be uncanny. African Grey Parrots have allowed researchers to address questions of whether bird sounds have specific meanings, similar to language in humans. African Grey Parrots are open-ended learners that learn calls throughout their lives. They are famously able to precisely mimic other sounds in their environment including human voices. What is truly remarkable about this ability is that they use stimulus specificity, which means that they can accurately match human labels to the corresponding items. For example, they can name the shape and colour of objects.

CALLS

Unlike song, calls are vocalisations produced by all birds, at all ages, during all times of the year for a variety of functions. Calls make up a much larger proportion of all bird vocalisations. Although we understand much less about calls than we do about song because research on

Male Superb Lyrebirds incorporate mimicry of other bird vocalisations into their displays. Females use mimicry in their vocalisations as well, though the female vocalisations may function more in territory and nest defence than courtship.

song has dominated since the 1950s and 1960s, recent research on calls has led to some interesting discoveries on how calls vary in function and development.

Calls are different from song because they tend to be shorter in length, have fewer components and are less complex than song. Calls are used by adults and juveniles as well as by males and females, but calls can exhibit sexual dimorphism – that is they are unique to each sex.

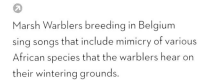 Marsh Warblers breeding in Belgium sing songs that include mimicry of various African species that the warblers hear on their wintering grounds.

Male Northern Mockingbirds use mimicry of other species in their songs more often during the breeding season than the non-breeding season, perhaps suggesting that females find mimicry attractive.

The vocal repertoire of the Common Chaffinch has been particularly well studied, and many aspects of avian vocal communication were first described in this species.

CALL REPERTOIRES

Some groups of birds exhibit much larger call repertoires than songbirds. It can be difficult to describe call repertoires because similar calls are used in varying contexts, so sometimes acoustic characteristics of different calls blend together. Some species, such as Chaffinches, have a well-defined repertoire of eight call types, which is a good representation of the ways in which birds use calls. Chaffinches have calls that are used by both sexes during all times of the year, such as a call used to indicate a threat (alarm call) and a call given during flight. Other call types are relegated to the breeding season. Some are used by territorial males when interacting with a mate. Females have calls they use to indicate willingness to copulate, and juveniles have calls they use when begging.

CALL DEVELOPMENT

Because research on the development of calls lags behind research on song, we know less about the influence of genes, learning, hormones and brain structure on call development and production. However, recent research on calls in a variety of species is starting to fill in the gaps. Traditionally, it was thought that calls were innate. However, laboratory studies suggest that some calls are learned or at least are flexible and modified by learning. In some species, such as the Yellow-naped Amazon, we observe that there are call dialects (regionally distinct call structures), which strongly suggests that learning is involved in producing some calls. Recent research on the contact calls of Green-rumped Parrotlets has shown that nestlings learn contact calls in the nest from their primary caregiver even if the caregiver is a foster parent.

There are many examples of calls that are sexually dimorphic. In Carolina Wrens, both sexes produce a harsh rasp when they are alarmed, but only males produce song. They also have sex-specific calls, including a chatter used only by females and often in response to or in conjunction with male song in a duet. As with song in songbirds, exposure to testosterone is necessary for learning and production of some calls that have been studied. Much remains to be learned about all aspects of call development and production, including the role of genes, learning, hormones and even the functions of calls. But the functional and developmental diversity of bird calls is a rich context within which to study the connection between the brain and communication.

Audio and video recording of Green-rumped Parrotlet nests allowed researchers to study how parrotlet calls are learned, and who the chicks copy.

Call learning in Yellow-naped Amazon parrots may help establish group identity: differences in calls allow other birds to know where individuals are from.

NON-VOCAL SOUNDS

Non-vocal sounds are an important component of acoustic communication in birds. Birds frequently use beaks and feet to produce percussive sounds, as well as the feathers of the wings and tail to make other non-vocal sounds.

Woodpeckers are famous for using their beaks to drum on resonant surfaces such as tree trunks, as well as other artificial surfaces. Like songs, the drumming patterns of woodpeckers are species-specific – that is woodpecker species can be recognised by the unique drum patterns they produce. Two species of North American woodpeckers – the Downy and Hairy Woodpeckers, which look very similar to each other – can be distinguished by their drumming patterns. The pitch of the drumming will depend on the substrate of the drum (whether it's a hollow tree, or the side of a rainspout, for example), but the tempo can be diagnostic. Interestingly, the smaller of the two species, the Downy Woodpecker, uses a slower tempo of about 15 taps per second while the Hairy Woodpecker uses a much faster tempo of about 25 taps per second. Some birds, such as Java Sparrows, make clicking noises with their beaks. Males will integrate clicks with song production, creating 'songs' that incorporate both vocal and non-vocal sounds. And some birds, such as White Storks, make more extensive percussive sounds by clapping their large mandibles together.

Feathers can produce sounds by percussion or by air passing through feathers. Some birds will use specialised feathers to produce whistling sounds during flight displays. These feathers are narrow and stiff so they vibrate at high frequencies when air passes through them during flight. For example, Common Snipe have two outer tail feathers that are adapted for producing sound during flight. Many species of hummingbirds also produce non-vocal whistling sounds during flight displays using their specialised tail feathers. The Club-winged Manakin can produce high-frequency whistles using stridulation of wing feathers, similar to the way that many insects produce sounds. Club-winged

Manakins are named for their highly specialised feathers that can generate sustained low-frequency tones, and research has shown that these tones also involve specialisations of one of the long bones of the wing. The ulna of the Club-winged Manakin is larger and more solid than the ulna of other birds, possibly to alleviate the stress of producing the sound or because it is directly involved in producing the sound. Having a larger, denser wing bone is unusual in birds because it is no longer optimised for flight. Thus, wing sounds produced by Club-winged Manakins are costly.

In addition, some birds use the percussion of wings or tail to produce non-vocal sounds without needing specialised feathers. During courtship displays, Red-Capped Manakins create snapping sounds by hitting their wings against their legs. Male Ruffed Grouse will use the feathers of their wings to produce a low-frequency drumming sound by beating their wings rapidly against the side of the body causing sound by compressing the air pocket between the wing and the body.

The distinctively shaped outer tail feathers of the male Anna's Hummingbird make a 'chirp' sound as the feathers flutter during display flights.

A low thumping sound echoing through a forest may indicate the presence of a Ruffed Grouse displaying from his favoured log or stump.

A Downy Woodpecker's drumming is used much like song during territorial competitions. Faster drums elicit stronger responses.

The bill clatter signal of White Storks is used in a variety of contexts, from courtship to nest defence.

The Red-capped Manakin's display involves colourful plumage, feather-snapping sounds and impressively fast footwork.

PLUMAGE

Feathers are unique features of birds and are not found on any other living vertebrate group. Some feathers have evolved to be stiff and rigid to aid flight, while other feathers are fluffy to help with thermoregulation, and there are also feathers which seem to serve no function other than display – but any feather can be colourful. The colours we see in feathers are produced by one or more of the mechanisms described below. Plumage colouration and pattern can provide camouflage but can also be used in communication.

PIGMENTATION

People are attracted to birds because they are highly visible and often exhibit an appealing array of plumage colours. Many of these vibrant colours, including the shades of red, orange, yellow and even some greens, as well as the subtler shades of black, grey, brown and rust, are all produced by pigments. Some pigments are generated by the birds' own physical processes, and some are obtained from food. Bright plumage is typically associated with males, but, in some species, the females have bright plumage – in the case of Red-necked Phalarope, females are brighter than males (reverse sexual dimorphism). Colourful plumage plays an important role in communication, as will be seen in Chapter 2.

MELANIN

Among animals, birds exhibit the highest degree of diversity of melanin pigmentation. Most bird plumage has at least some melanin pigment because melanin makes feathers stronger by increasing resistance to wear and feather-eating bacteria. Melanin is found in two varieties: eumelanins that produce darker colours such as black, brown and grey; and phaeomelanins that produce warm, earthy tones including tan, reddish or light brown. Melanin granules are produced in cells of the epidermis called melanocytes that occur near the feather follicle and are deposited into the developing feather. The pathways for melanin production in birds are encoded by numerous genes, and variants in genes such as MCR1 (melanincortin receptor 1) lead to variation in the presence of melanin in plumage.

CAROTENOIDS

Carotenoids are pigments that are commonly found in nature. In birds, as in other organisms, carotenoids produce bright red, orange and yellow colours. Birds obtain carotenoids from the foods they eat, which are modified in many ways to become the pigments deposited into developing feathers, but genes are involved in the assimilation of those carotenoids into the coloured feathers. Carotenoids are found in most plants and many invertebrates.

PSITTACOFULVINS

A notable exception to the assimilation of carotenoids into feathers to produce red, orange or yellow plumage can be found in parrots (the order Psittaciformes), which use pigments called psittacofulvins to produce these colours. It has been known for a long time that parrots use these unique pigments to produce colourful plumage. However, it was not known until recently that, unlike carotenoids, psittacofulvins are not acquired from food and are likely produced by specialised cells, such as melanocytes, near the feather follicle. Very recently a gene involved in the production of psittacofulvins was discovered.

In this pair of Red-necked Phalaropes, the brighter bird at the top is the female. This reversed sexual dimorphism is associated with a breeding system in which females compete to attract males, while males incubate the eggs and provide the parental care.

PLUMAGE   35

PORPHYRINS

Porphyrins are iron-containing pigments, including haemoglobin and liver bile pigments. They produce reddish or brown feathers in some groups of birds, including owls. These pigments are not as chemically stable as melanins or carotenoids and degrade in sunlight, so they are present only in new feathers. A couple of related porphyrin pigments contain copper instead of iron – turacin which produces bright magenta in the wings of Turacos, and turocoverdin, which produces green in a few species, including Turacos. Green is more commonly produced in birds by a combination of structural colour and pigments (see below).

WHITE

White colouration occurs when plumage lacks pigment and the arrangement of keratin molecules causes the scattering of all visible wavelengths of light (known as incoherent scattering). White is used in communication (see Chapter 2), and is also important for camouflage, and also possibly for thermoregulation because white reflects rather than absorbs heat. White plumage is commonly found in Arctic bird species such as Ptarmigan and probably functions as camouflage. White plumage is also commonly found in waterbirds, such as shorebirds, in which it is thought to act as camouflage by providing countershading. The combination of dark dorsal plumage with white ventral plumage reduces the visibility of objects that are lit from above and may also be important for thermoregulation.

There are cases of albinism in birds in which a genetic mutation results in a lack of pigment in some or all feathers. In these cases, the feathers appear to be white, but the keratin proteins that make up feathers are transparent, and so the white colour we see in such feathers lacking pigment is the result of the way the underlying structures of feathers scatter light.

↑
This Fischer's Turaco illustrates the unusual pigment-based green and magenta colours found primarily in turacos. The lower photo shows the magenta feathers in close up.

→
The winter plumage of this Willow Ptarmigan lacks pigment and scatters light to produce white colouration. In summer, the ptarmigan will be clad in red, brown and black, using pigments to achieve the same goal of camouflage.

These two views of a Ruby-throated Hummingbird demonstrate the directionality of iridescent feather displays.

STRUCTURE

Bright blue or shimmering, iridescent plumage is not the result of pigments but rather the underlying structure of the feathers. The arrangement and density of the keratin molecules – the microstructure – that make up the feathers can result in a scattering of reflected light from the feathers. This appears as a bright or a shimmering patch of colour.

IRIDESCENSE

When melanin granules are arranged in a regular layer as opposed to scattered throughout the feather structure, light can be refracted as through a prism (known as coherent scattering). The colour that is emphasised is affected by the arrangement of the keratin molecules in the feather. But the visibility of the colour is also dependent on the angle of the viewer. Iridescent plumage, such as that of male Ruby-throated Hummingbirds, can appear bright shimmering red when viewed from the appropriate angle. But when viewed from another angle, the feathers look black because of the underlying melanin granules.

BLUE-ULTRAVIOLET

Coherent scattering also produces most blue plumage. However, the lack of organised melanin granules will result in scattering of light that is independent of the

The blue plumage of the Eastern Bluebird is produced by light scattering and contains UV wavelengths visible to other birds but not to humans.

This male Varied Bunting shows how red carotenoid pigment and blue structural colours can combine to produce purple colouration.

angle of the observer. In blue plumage, the keratin molecules are arranged so that light is only reflected in the blue part of the electromagnetic spectrum. Variation in blue plumage, such as the bright blue in male Eastern Bluebirds and the duller blue in females, corresponds to the thickness of the keratin layers.

GREEN AND PURPLE

Green and purple plumage can be produced by pigments alone or by a combination of pigments and structural colour. Wild Budgerigars are green because of a combination of yellow pigment (a psittacofulvin) and a structurally produced blue. Genetic mutations in captive Budgerigars produce a wide variety of colour variants. One mutation is known to interrupt the psittacofulvin deposition and individuals with that mutation are blue – only the structural colouration is present. Other mutations interrupt melanin deposition preventing light-scattering to produce yellow colouration. Combinations of approximately thirty mutations in Budgerigars result in a wide variety of colouration from white (albinism) to shades of grey, green, yellow and blue.

If a structural blue is combined with a red pigment, the resulting plumage colour is violet or purple, as seen in the male Varied Bunting.

OLFACTION

One of the communication channels in birds that is less well understood is olfaction. This is a relatively new area of study and is not as easily observable in birds as song or colour. In fact, you would probably never detect an odour from a bird because many birds do not have a smell. One interesting exception is the tangerine smell emitted by the Crested Auklet, a colonial seabird of the Pacific northwest, which is detectable to anybody within 1km of a colony.

Further, because birds do not actively 'sniff' like mammals, it is likely that birds communicating with olfaction are not observed in the same way as they are when using colour and sound. Until recently, many researchers dismissed the possibility of birds using olfaction in communication because most people

thought birds had a poorly developed sense of smell, with a few notable exceptions, such as the well-known olfaction ability of kiwis, vultures and seabirds (particularly shearwaters, petrels and albatrosses). However, recent research has discovered that not only do most birds have a keen sense of smell, but they are likely using olfaction in communication. Some have suggested that the bill wiping commonly observed in birds functions in chemical communication by spreading preen oils onto surfaces.

The anatomy of olfaction in birds is similar to the anatomy found in other terrestrial vertebrates that are known to have a keen sense of smell in that birds possess specialised sensory receptors that are sensitive to chemical stimulation, together with neurons that project into the olfactory bulbs of the brain. But there remains a lot of variation in birds' anatomy that suggests that there is considerable variation across species in terms of the perceptual abilities of smell. Nonetheless, it is likely that most birds are capable of at least some olfaction. Many studies show that olfaction in birds is important for finding resources and detecting prey or, in the case of colonially nesting seabirds, for relocating nests and burrows in a large colony. Research is also accumulating that shows birds can produce chemical cues which are detectable by the same species and used in communication.

There is likely to be a primary difference in the way birds use chemical communication when compared to other vertebrates with well-known capacities for chemical communication, such as mammals. Mammals – including canids, felines and primates, among others – detect chemicals used in communication (pheromones) with a specialised sensory receptor, the vomeronasal organ. Animals of the same species (known as conspecifics) can decode the chemical signatures in pheromones to acquire important information such as an individual's identity, health, reproductive condition and the location of territory boundaries. Currently, there is no evidence

Crested Auklets
communicate with vocal
signals, visual signals, physical
displays and odours.

Seabirds like this Buller's
Shearwater may use olfaction
both for navigation and for
finding food.

to suggest that birds have a functional vomeronasal organ and instead are likely using olfaction in chemical communication.

The main source of smell produced by birds comes from the uropygial gland located at the base of the tail on the dorsal side. Birds use oils produced by the gland for preening. Preening typically involves birds collecting a bit of oil secreted by the uropygial gland with its beak and then using the beak to spread the oils across the feathers while simultaneously 'zipping up' the barbs and barbules to maintain the structural integrity of the feathers. By spreading oils, birds are conditioning the feathers to keep them clean and supple and in good condition, which is especially important for flying and swimming, and thus survival.

Chemical analyses on a growing number of species show that the substances that make up the oils (fatty acids) vary in their chemical properties between species, sexes and seasons. Thus, at the very least, it is likely that uropygial secretions can contain information about species' identity, sex and breeding stage. In a study on Mallard ducks, researchers prevented some males from perceiving smell by surgically severing olfactory nerves. When compared to males that had similar surgery without severing the nerve, males with the deactivated sense of smell did not perform the usual social behaviours associated with breeding such as mounting females and copulation. The manipulated males appeared to behave normally around other males, using aggression, but failed to behave normally around females. Thus, the surgery appeared to impact their ability to detect olfactory cues from females that would lead to normal reproductive behaviour.

Experimental studies on European storm petrels, a colonially nesting, highly philopatric species, found that adults can discriminate kin from non-kin. This would be desirable in this species because it allows adults to avoid inbreeding and the possible negative consequence to fitness. In a recent study in the hybrid zone between Carolina and Black-capped Chickadees, researchers have discovered that adults are using olfaction to avoid mating with the wrong species of chickadee. Being able to avoid the mistake of mating with a closely related species in chickadee hybrid zone is important because hybrid offspring have lower rates of survival than purebred offspring.

Though most odorants in birds are produced by the uropygial gland, the tangerine smell of the Crested Auklet is not. It is thought that both sexes in Crested Auklets produce this odour in feathers located on their necks during the breeding season. The odour may be important for courtship when the adult birds perform a display that involves neck rubbing – here adults will actively 'sniff' each other's necks by burying their beaks deep in the neck feathers. There is also some evidence to suggest that the chemical compounds found in the feathers may serve as insect repellent. Thus, higher quantities of odorant may indicate the ability of adults to repel ectoparasites. More detailed studies on breeding adults find that the quantity of odorant is correlated with crest size in males and with immune function in both males and females. Thus, by sniffing each other's necks, adult Crested Auklets may be acquiring information about immunocompetence. Since it is at least partially heritable, a mate with a more robust immune system is likely to ensure higher levels of fitness in offspring.

European Storm Petrels are more attracted to the scent of unrelated individuals.

Olfaction clearly seems to play a role in Mallard sexual behaviour, but the source of the presumably attractive odour is unknown.

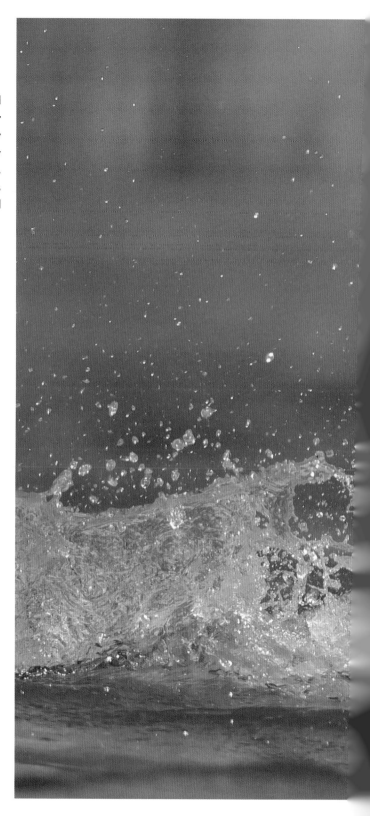

MOVEMENT DISPLAYS

In addition to all the modes of communication discussed so far, birds can combine them with dances and other displays into a multimodal performance. Display behaviour is a ritualised set of actions that is typically repeated intact from one performance to the next. Many of the examples of non-vocal sounds using feathers during flight are used in combination with a ritualised flight performance that draws attention to the display.

The Red-capped Manakin will use whistles and claps from specialised feathers while simultaneously performing a moon-walk-like dance to attract the attention of nearby females. Male peacocks will maximise visibility of bright plumes by vibrating them in the presence of females. Males can perform in solitary dances or they can perform together in groups called leks. Sometimes males perform a coordinated display to attract the attention of the female. In Lance-tailed Manakins, two males perform a coordinated display to a female, who may mate with one of them.

Males and females in some species coordinate and perform dances and songs together. Western Grebes perform attention-grabbing displays as a solo dance, in pairs of male and male, male and female, and in sets of multiple males and females. The dual display of Western Grebes varies from a coordinated bobbing and raising of crests to rising out of the water from a resting position to run across the surface at high speeds while holding their necks in a curved posture with crest feathers raised. The display is accompanied by vocalisations and serves multiple functions, including aggression, mating and pair bonding. Many bird species perform displays that involve plumage, movement and vocalisation, but only one species we know of uses a combination of all the modes of communication plus olfaction – the Crested Auklet.

For Western Grebes, synchronised swimming can help to strengthen the pair bond.

MALE–FEMALE COMMUNICATION

Male-female communication in birds generates the conditions that produce the most elaborate displays found in animals. The evolution of colourful plumages, complex vocalisations and exhilarating behavioural displays in birds help promote the fitness of each partner, because such traits provide a means of communicating information that is essential for maximising reproductive success.

FAVOURED TRAITS

Charles Darwin was the first to suggest sexual selection as the main force acting on birds to produce colourful plumages and melodic songs, and many researchers have since supported his hypothesis. Darwin proposed that sexual selection favours traits that improve the quality or quantity of reproductive opportunities. The male traits favoured by sexual selection tend to be exaggerated morphological or behavioural traits that are used in competition with other males either directly, such as the spurs on the legs of male wild turkeys, or indirectly by advertising for females, such as the elaborate plumage in male peacocks.

Occasionally, sexual selection favours exaggerated morphological traits in females, too. But the trait favoured most often by sexual selection in females is female mating preferences, which can be the driving force in the evolution of exaggerated traits in males. In birds, sexual selection results in some of the most extreme forms found in any animals. Most notably, birds are known for their bright, flashy plumage and wide range of vocalisations including complex, melodic songs. Bright plumage and beautiful songs are traits that give males an advantage in accessing females either through direct competition with other males or by being the most attractive to females.

➔

During displays, the male Indian Peafowl will vibrate his elaborately ornamented plumage to attract the attention of females.

FEMALE MATING PREFERENCES

The potential for sexual selection is greater for males than for females because males are less physiologically constrained to produce offspring and can maximise reproductive success by attracting as many mates as possible. Therefore males tend to be competitive for access to females, whereas females tend to be less competitive and more choosy about mates. Males can directly compete for access to females through male–male interactions, often involving violence.

Traits that give males an edge in exclusive access to females through direct interactions with other males are favoured by intra-sexual selection. When males cannot monopolise the attention of females through male–male interactions, the only alternative is to advertise for females. And, much like the flashy and catchy advertisements designed to attract your attention, male birds will use flashy plumage displays and catchy songs to attract the attention and favour of females. Males who can produce the advertisement that is most attractive to females will benefit by having more opportunities for reproduction and the most offspring. Exaggerated male traits that give them an advantage in attracting female attention are favoured by inter-sexual selection.

Males with exaggerated traits benefit by acquiring mating opportunities through being attractive to females. But why should females have preferences for males with those exaggerated traits, especially when attending to male advertisements can be costly to females in the time spent searching for preferred mates, increased exposure to predation and disease, and potentially foregoing mating opportunities if they cannot find a male that matches their preference?

Sexual selection theory predicts that female preferences are favoured because there are benefits to females. In general, females tend to be choosy because they are physically and physiologically limited in how many offspring they can have. The best way for females to maximise reproductive success is by mating with the highest-quality male possible so that she can have

high-quality offspring. High-quality mates can benefit female reproductive success by providing good resources, helping to rear young and successful genes for their offspring. In such cases, any preference that results in mating with a high-quality male will be favoured by inter-sexual selection. When it is difficult for females to directly assess male quality, inter-sexual selection should favour traits in males that are correlated to male quality, such as exaggerated behavioural or morphological traits.

Demonstrating the influence of female mating preferences on the evolution of exaggerated male traits is theoretically important but empirically challenging – how can researchers determine which males or male traits female birds prefer? Correlative field studies can

provide important insights. Males with putative preferred traits may pair earlier or attract more females to their territories. But we cannot know for sure that female preferences for the trait of interest are driving the patterns. For example, if a male trait is influencing his ability to defend a high-quality territory, then patterns of settlement and nesting by females may reflect an attraction to the territory and not to the males themselves. Therefore, experimental studies in both the field and the laboratory can help provide more direct evidence of female mating preferences that corroborate correlative studies. A good experimental study will reduce or eliminate confounding variables when assessing mating preferences in females.

Western Capercaillie male displays to numerous females simultaneously at a lek. Females are much smaller than males, making Capercaillies one of the more extreme examples of sexual dimorphism in extant birds.

SONG

Females are attracted to male song, the presence of which stimulates reproductive activity in them. Research on ring doves in the 1960s demonstrated that playing recordings of male song would induce reproductive behaviour in females, such as nest building and egg laying. In studies on European Starlings, House Wrens and European Pied and Collared Flycatchers (all cavity nesters), researchers undertook experiments in which they played male song from some nest boxes and not others, and then observed in which boxes females built nests. They found that females were attracted to and built nests more quickly in boxes in which male song was being played than in boxes without males or songs.

Clearly, song is attractive to females, but how can we discover to which features of song females are reacting? Are some songs or styles of singing more attractive than others? Since song cannot be easily manipulated in live males, most studies of female preferences use recorded songs, not live males. In 1977, it was discovered that hand-raised female Brown-headed Cowbirds would perform a solicitation posture in the laboratory in response to male song. Solicitation postures are performed by wild females when they are receptive to male attempts to mate. However, subsequent studies did not observe wild adult females perform solicitation in the laboratory. In 1981, researchers discovered that female

House Wrens are sexually monomorphic. Males fill multiple cavities with sticks and the female chooses one of the cavities in which to put her nest.

Male European Pied Flycatchers exhibit striking black and white plumage. There is evidence to suggest that the size of white patches on wings and forehead are favoured by sexual selection.

Brown-headed Cowbirds are famous for being obligate brood parasites. Females do not build their own nests but lay their eggs in the nests of other species.

songbirds in the lab would perform solicitation displays to male song if they were primed with oestradiol, a hormone important in reproductive behaviour that is elevated in females during the breeding season.

In solicitation assays, preference is measured by the number and quality of solicitation postures females perform during song playback – solicitation is a clear sign of preference in wild females. However, solicitation assays require oestrogen implantation, which involves minor surgery and doesn't always work in every species.

Other experimental approaches have been developed to assess preferences for song. In operant conditioning, females are trained in the lab to elicit a song through perches that trigger playback. The idea is that a female will most often trigger songs that she prefers. The development of experimental methods for measuring preferences for song have led to a better understanding of how song serves as an important signal to females when choosing mates.

SONG AS A SIGNAL OF QUALITY

The advertisement songs of male songbirds have proven to be especially attractive to females. Males sing most during the breeding season with periods of peak singing corresponding with periods of peak fertility in females. We expect that songs are attractive to females because they provide information about male quality. However, it turns out that song is not particularly expensive to produce. Studies on oxygen consumption of males during singing do not reveal any significant metabolic costs in song production. So, how does song serve as a signal of male quality to females? Song is not a single trait but a suite of traits – there is potential for multiple aspects of song to provide reliable information to females. There are several ways in which song has proven to be costly and as such only the highest quality males can produce the most exaggerated version.

SONG QUANTITY

Even though the metabolic costs of singing for males is marginal, the time spent singing can exclude participating in other activities such as self-maintenance or caring for young. Song quantity, including song duration and song rate, has been studied as a feature used by females to assess male quality. In a study on free-living European Starlings, a cavity-nester species, researchers set up nest boxes for males and recorded songs. They found that males who sang longer song bouts had more females visiting their nest boxes, paired earlier and had more young. Although the researchers did not directly assess females' responses, these studies suggest that longer song bouts are more attractive to females.

In a follow-up study using an operant-conditioning approach, researchers found experimental evidence that females preferred longer song bouts. Further research revealed that males who sang longer song bouts also had better immune systems. In another study, in which nutritional stress was manipulated during the nestling period, researchers found that stressed nestlings had lower immunocompetence and became adults that sang shorter song bouts. So, by having preferences for males with longer song bouts, females were mating with males that had better immune systems or experienced less stress during development. Because immunity is at least in part the product of genes, by mating with the males that have superior immunity, females may be ensuring that their offspring will be primed for fighting disease and can live longer.

⊘

Male Common Starlings use complex songs to establish nest sites and attract mates.

⊘

Ovenbirds are so named because females build domed nests that look like little ovens on the forest floor.

Unlike most bird species, Common Nightingales sing throughout the night as well as the day.

Male Sedge Warblers start singing within just a few hours of arriving on the breeding grounds.

SONG QUALITY

Repertoire

Song quality is a feature of song that is a bit more nebulous to define than song quantity. One feature of song that is commonly used as indication of song quality is song complexity. Songs that are sung with more complexity are thought to be a more extreme version of the song, akin to an acoustic peacock tail that has a diversity of eyespots. Complexity is typically expressed as the number of different song types males sing—song repertoire—or the number of different elements males use to construct songs—element repertoire.

Song repertoires can vary from one main song type, as in Ovenbirds, to hundreds of song types, as in Nightingales, and again to many hundreds of song types, as in Brown Thrashers. Element repertoires have a similar range of variation across species to that of song repertoires. Some species, such as Great Tits, have a small element repertoire. Others, such as Sedge Warblers, have a larger element repertoires of seventy-five elements which they recombine for a potentially infinite variety of song types. More importantly, however, for our discussion of how repertoire size is a signal of quality, is that there is variation within species.

For example, repertoire size in male Song Sparrows ranges from five to fifteen song types. The idea is that males that use larger song or element repertoires are exhibiting more complex and therefore higher-quality song than males that use smaller repertoires. And, like many other species tested, female Song Sparrows demonstrate preferences for larger repertoire sizes. In the lab, females perform more copulation solicitations to playback with eight song types than to playback with four song types. And, in the wild, male Song Sparrows that have larger repertoires produce more offspring in their lifetimes. But what do females get by mating with males with more complex songs?

Zebra Finch

SONG REPERTOIRE SIZE AND BRAIN COMPARED

The figures show cross-sectional views of Zebra Finch brains, and highlight the areas associated with song. Male song is learned and involves many centres of the brain for both song learning and production. Some of the areas associated with song learning and production are present in females but are much smaller than in males or are missing (area X). Researchers hypothesise that song learning and development can reveal how well males are able to withstand periods of stress, and thus song can serve as a signal of quality to females. Measures of song quality, such as repertoire size, have been shown to be positively correlated with the volume of the area of the brain associated with song (HVC). Experiments that stress males during development have shown a reduction in some brain areas that are associated with song (HVC and RA).

SONG CONTROL NUCLEI OF THE BIRD BRAIN:
HVC: High vocal centre **MAN:** Lateral magnocellular nucleus of the nidopallium **RA:** Robust archistriatal nucleus **X:** Area X **ICo:** Intercollicular nucleus **nXIIts:** Hypoglossal nucleus

Neuronal pathways
The arrows indicate the pathways in the brain for generating and learning song. Some of the pathways are important in producing song and sending the signal to the syrinx (motor pathways). Other pathways are important for song learning.

↑ ↗

Laboratory studies in domesticated birds (Canary, left) as well as research on wild bird species (Great Reed Warbler in Europe, centre, and White-crowned Sparrow in North America, right) have contributed significantly to our understanding of how bird song is used in communication.

Both repertoire song and element size have been implicated as signals of quality because male songbirds must learn song. Learning a larger repertoire requires males to possess and invest in the brain structures associated with song learning and production. Birds learn song during the sensitive period that corresponds to a brief period when recently fledged juveniles are learning how to be independent, typically a stressful time for young birds. Therefore the quality of song learning may be indicative of the overall condition of males as well as a genetic ability to withstand periods of stress.

In laboratory studies on brain growth and development under stress, researchers have found that nutritionally stressed Song Sparrows, European Starlings and Zebra Finches all have smaller volumes of nuclei in the brain associated with song production (HVC and RA) compared with controls, and that song production was negatively impacted. Stressed Song Sparrows produced poor copies of tutor songs and stressed Zebra Finches produced simpler songs.

In a laboratory study on a different cause of stress, Canaries infected with malaria during song learning produced a smaller repertoire size as adults than adults

that had not been infected. In the cases of the effects of stress on song quality in Song Sparrows, Zebra Finches and Canaries, females in the lab discriminated against songs produced by stressed males. In studies on Great Reed Warblers, researchers accumulated evidence across many studies demonstrating that females have preferences for more complex song repertoires, and that males that have more complex song repertoires produce more viable offspring. So, by mating with males that have higher-quality songs, female Great Reed Warblers are choosing high-quality males which enable them to produce young better able to survive and reproduce.

Local song types

Song learning has another interesting consequence on the quality of bird song. Random errors in learning can lead to locally distinct song structure or dialects. Songs within species will vary, often subtly, from one locale to another. Geographic variation in song can be abrupt with discrete boundaries, as observed in White-crowned Sparrows and females will demonstrate a preference for local song types in laboratory studies. Song structure can also vary gradually with distance as observed in Song Sparrow song. Despite gradual and subtle differences in song structure, female Song Sparrows will begin to discriminate against songs recorded from as near as 34km. Male discrimination in Song Sparrows does not occur until males hear songs from at least 540km away. Females are therefore more sensitive to geographic variation in song and, this too, can be a way for females to assess male quality.

In song learning experiments with Song Sparrows, young males that were nutritionally stressed produced less accurate and precise versions of local songs than did control males. Wild-caught, adult females demonstrated a preference for the songs produced by control males when compared to songs produced by stressed males. Females can therefore assess song quality as the truest rendition of the local song types and, as with song complexity, this can indicate male quality because males have invested in learning and producing local song.

Song performance

How well males perform songs is becoming recognised as another feature of song quality that females desire, possibly because it is another indicator of male quality. Even though singing is not energetically expensive, precise coordination of the vocal and respiratory tracts during song to produce certain kinds of notes, or to maintain high amplitudes across the song, is physically challenging. Like performing opera, some types of singing in birds requires males to overcome physical or physiological constraints and only some males can sing songs that push the physical boundaries. Thus, only the highest-quality males should be able to sing the most challenging song types.

Swamp Sparrows are the subjects of many studies on bird song in the US, including the first ground-breaking research that showed that songbirds learn their songs rather than 'know' them instinctively.

Trill rate and frequency bandwith of sparrow songs
A study of songs from 32 species of sparrows plotted trill rate against frequency bandwidth. The results support the idea that there is a physical trade-off in song production, such that it is impossible for males to sing songs that have both the fastest trill rate and the broadest frequency bandwidth. Thus, songs that are produced at the physical maximum (along the boundary of the trade-off) are more difficult for males to produce (high performance) than songs that are further away from the boundary (low performance). Females have been shown to demonstrate preferences for high-performance songs. Redrawn from Podos (1997).

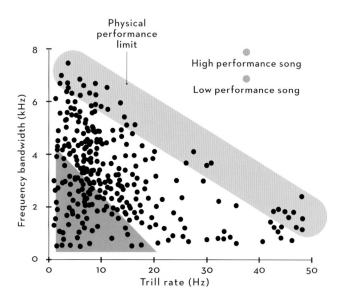

In the laboratory, female Swamp Sparrows and Canaries demonstrate preferences for trilled songs that simultaneously have broad frequency bandwidth and rapid trill rate, a combination of traits that should be physically challenging to achieve. Female Dusky Warblers prefer males that can maintain high amplitudes during song. In addition, male Dusky Warblers that sang higher-performance songs lived longer. Females with preferences for higher-performance songs may have been providing good genes for the offspring that would improve longevity.

If you watch a bird sing, you will notice that the bill moves in concert with the frequency of sound being produced. During song, birds are rapidly coordinating bill movements with respiration. Trilled songs that include notes repeated in rapid succession present a challenge for birds. Birds produce sound through a specialised organ called a syrinx (see pages 20–1), a modified tracheal ring that has enervated membranes that vibrate at certain frequencies. Once the sound is produced by the syrinx at the base of the trachea, birds use the trachea and beak to filter out harmonic overtones to produce pure tone notes. A bill that is most open favours high-frequency sounds and a bill

⬆
Dusky Warblers are migratory birds of Eurasia that winter in Southeast Asia and breed in the Arctic, where males defend their territories with a loud, simple, trilled song.

that is most closed favours low-frequency sounds. So, during rapid trills, males are constrained to reach the broadest-frequency bandwidths, and at the broadest-frequency bandwidths males are constrained in producing the most rapid trills.

Songs that are closest to the physical boundary are more physically challenging for males than songs that are further from the boundary. In Swamp Sparrows, size and age are correlated with performance. So females with preferences for high-performance songs are attracted to the larger males that are likely to obtain higher-quality territories and possibly good genes for their offspring which improve longevity.

PLUMAGE

In 1990, a comprehensive set of experiments on female mating preferences showed that, when given a choice between males that ranged in plumage colour from dull yellow to bright red, female House Finches strongly preferred males with the brightest red plumage. The experiments presented female finches with males that varied in plumage colour that was manipulated to mimic the natural range of colour variation seen in the wild. Female preferences were measured by the amount of time they associated with males.

Manipulating male plumage experimentally is necessary to control for other male characteristics that females might prefer and that might correlate with bright red plumage. In one experiment, plumage was manipulated by using hair dye to brighten plumage, and in another experiment, plumage was manipulated by altering the availability of carotenoids – the pigment responsible for the colour red – to captive males. No matter whether males were presented to females individually, in a group, in natural or manipulated forms, females always demonstrated a preference for bright red males.

Since this groundbreaking study on House Finches, most studies of female mating preferences for plumage characteristics take advantage of the relative ease of manipulating plumage on live males and measuring female preferences for plumage traits displayed in them. In this way, female mating preferences for plumage displays can be addressed in the laboratory and in free-living populations. In free-living birds, plumage characteristics of males can be manipulated, and female preference can be measured by relative timing of mating or, in polygynous species, the number of females that settle in a territory.

In a study on a free-living population of Red-collared Widowbirds, researchers addressed mating preferences of females for long tail plumes by manipulating the tail length of males and measuring how many females nested on the territories of males with differing tail lengths. Males that were experimentally manipulated to have longer tails had significantly more females nesting on their territories than males that were manipulated to have shorter tails. Another important finding from this study was that males with shorter tails could defend territories just as well as males with longer tails. Therefore, the researchers concluded that long tail plumes in males were favoured by female mating preferences and not by male–male competition.

Variation in the brightness and extent of red colouration in male House Finches is correlated with nutritional condition. Females demonstrate a preference for the reddest males.

Male Red-collared Widowbirds perform conspicuous flight displays on their territories to attract the attention of females.

PLUMAGE AS A SIGNAL OF QUALITY

At the most basic level, if males want to advertise to females using visual signals, they must produce signals that females can perceive and are easy to detect. Males can invest in growing specialised plumes used only in display that require more resources to build and stand out to females, such as the exaggerated plumage displays produced by male wild Turkeys. Males can also invest in more colourful plumage. Because females can perceive visible wavelengths as well as ultraviolet (UV) wavelengths, males produce colour signals in the range of visible and UV colouration. But not all colours are created equally! Some colours are more costly to produce or maintain than others and those are the colours we tend to see being used by males to attract the attention of females. For example, melanin, which produces black, brown and rusty plumage colouration, is produced endogenously by birds, making these colours relatively less expensive than those produced by carotenoids. Carotenoids produce yellow, orange and red plumage and must be acquired from the environment, which is costlier. Costly signals are the best at providing females with reliable information about male quality because not all males will be able to afford the costs associated with producing those signals.

Wild Turkey males perform displays to females gathered in clearings by slowly strutting around with fanned tails and puffed-out plumage.

CAROTENOID COLOURS

Rigorous and extensive studies on House Finches have shown that the degree of redness expressed in male plumage is the result of both access to carotenoids and good nutrition. It is not only costly to obtain carotenoids in the diet, it requires that males are in good condition to be able to convert diet-based carotenoids into the kinds of carotenoids that can be deposited into growing plumage.

In American Goldfinches, males experimentally infected with a gut parasite produced carotenoid-based plumage that was less bright than in control birds. In an observational study on a disease outbreak in House Finches, researchers found that males were disproportionately affected by the outbreak and that survivors had redder plumage. However, the expression of melanin pigment in those same infected birds was unaffected. Therefore colours that are produced by pigments obtained from the diet – such as red, orange and yellow (see page 34) – appear to be associated with mate choice because they are often condition-dependent.

Female House Finches that choose males based on red plumage will be mating with males that are in better condition and may be more resistant to disease. In contrast, colours that are produced by melanin pigments, which are less dependent on condition, are more often associated with signals of status. However, studies on Eastern Bluebirds have demonstrated that males with larger and brighter eumelanin patches provisioned young more than males with smaller patches. Eumelanin produces a chestnut colouration and may also be used in male–male interactions to gain access to high-quality territories.

American Goldfinches are striking and strange little birds. They are one of the latest breeding songbirds in North America because they wait until seeds are produced by thistles and milkweeds.

Blue-black Grassquits are native to South and Central America and are commonly found in open grasslands and fields.

Blue Grosbeaks are handsome, large North American members of the cardinal family, which breed in open fields across the southern US and Great Plains. Adult males do not attain their full adult plumage until after their second breeding year.

STRUCTURAL COLOURS

Some plumage colours, including white, green, blue, as well as iridescence, are produced by feather structure, rather than by pigments (see pages 38–9), and so do not require special diets. However, the development of feathers that are green, blue or iridescent requires a precise arrangement of biomolecules that has been shown to be condition-dependent.

Feather structure can also produce colour that reflects in the ultraviolet (UV) part of the electromagnetic spectrum. And, until recently, researchers did not know that plumage colour in birds contained hidden markings. Some of the first studies on UV perception in birds showed that, when given a choice between individuals that are viewed in natural light and individuals viewed through a UV-blocking filter, both males and females preferred partners viewed in natural lighting conditions, reflecting UV – suggesting that they need to be able to see the hidden UV markings to make a choice of partner. Spectrometers that can measure colour reflectance into the UV have since confirmed that many birds have plumage that reflects in the UV range.

In studies of Blue Grosbeaks and Blue-black Grassquits, feather growth rate – an overall sign of nutritional condition during moult – has been shown to correlate with blue plumage colour and brightness. Males that are in better nutritional condition, and can grow feathers faster, also have brighter, more blue (UV-shifted colour) plumage. In Blue Grosbeaks, brighter males also fed their chicks more than less bright males. Similar results have been found in Eastern Bluebirds – brighter males provisioned more, had larger nestlings and fledged more young. Despite the obvious benefits to females of choosing bluer males in some species, and studies showing the presence of UV reflectance being important to mate choice, experimental studies have been unable to demonstrate preferences by females for natural variation in blue plumage.

Peacocks are often used as an icon for sexual selection – males have exaggerated plumage, including a long train with iridescent tips, the 'eyespots', which they vibrate in the presence of females. Researchers have observed that mating success at leks is positively correlated with the number of eyespots. Other studies have also shown the males with more eyespots tend to be older, so males that have the most eyespots are older and preferred by females. Females that mate with males having the most eyespots lay more eggs and their offspring are more likely to survive. So, by having more iridescent eyespots, males acquire more mating opportunities, which lead to higher reproductive success. And, by having preferences for males with more iridescent eyespots, females are laying more eggs and have young with higher rates of survival.

WHITE PLUMAGE

Research on white plumage as a signal between males and females lags behind research on other plumage colours, possibly because of the assumption that white plumage is a default plumage colour and that is the product of a lack of melanin or structure to make plumage colourful. However, recent studies suggest the white plumage patches are condition-dependent and may be an important signal in male-female communication. Experimental studies on Dark-eyed Juncos which are induced to grow new tail feathers under nutritional stress and in a nutrionally enriched environment found that the individuals that were stressed grew tail feathers with duller white patches than those in a nutritionally enriched environment. These results suggest the brightness of white plumage patches in Dark-eyed Juncos are condition-dependent.

Although there is clear evidence that brightness of tail feathers is used in male–female communication in Dark-eyed Juncos, there is evidence that brightness of white patches is an important signal in Black-capped Chickadees. Researchers observed that males with brighter white plumage patches suffered less extra-pair paternity in their nests and thus had higher reproductive success than males with duller white plumage patches. Therefore white plumage patches may be condition-dependent and used by females when making decisions about mate choice.

The Dark-eyed Junco is found throughout North America, breeding in the north and wintering in the south. It exhibits enormous geographic variation in plumage colour but can always be identified by the flash of white outer tail feathers.

Male and female Black-capped Chickadees have similar plumage, making it difficult to tell the sexes apart in the field. Both males and females use the 'chick-a-dee' call, but only males sing.

DISPLAYS AND DANCES

Male birds can enhance acoustic and plumage signals, and improve their visibility, by performing conspicuous displays and dances. These displays can also act as signals. Multimodal displays that increase conspicuousness are especially prevalent in lek mating systems. These occur across several species of birds, and usually involve males performing displays that involve visual, acoustic and behavioural displays.

⬇

The Sage Grouse is the largest grouse in North America and is currently of conservation concern due to threats to its sagebrush habitat in the American west.

On leks, males congregate during the breeding season to perform displays for visiting females. Males arrive at the lek site and set up shop by staking out a small piece of ground upon which to perform attention-grabbing displays, usually in the clear sight of other males. A female visiting the lek observes males, mates with the male whose performance she likes best and then goes off on her own to raise young without any contribution from the male. For example, male Sage Grouse congregate on leks in the sage brush grasslands of North America, where they perform dances that involve plumage displays, vocalisations and booming percussive sounds made by inflating visible throat pouches. At a lek, males are highly visible and audible, and females will observe many male performances simultaneously.

Some lekking species will modify the surrounding habitat to be more visible. In Jackson's Widowbirds, which breed in east African grasslands, males on leks create stage arenas by flattening the grass into a ring with a central tuft. Once the stage is set, males will perform jumping displays which accentuate their long tail plumes while making vocalisations. Females will visit and mate more often with the males that have longer tail plumes and greater jump rates. Greater display rate attracts more females to the lek, and males with the longest plumes copulate with more females. Tail length is related to male condition. So, by attending to male displays at a lek, females can mate with high-quality male Jackson's Widowbirds.

Among the most extraordinary examples of birds that perform behavioural displays on carefully remodelled stages are Bowerbirds. Bowerbirds are a group of birds that occur in Australia and New Guinea and perform displays in structures that they build and decorate, called bowers. Bowers range from an ornamented clearing, like the stage arena of Jackson's Widowbirds, to elaborate structures that are decorated with a colourful array of items which can include feathers, bugs and berries, as well as human refuse like yoghurt lids and bottle caps. Some studies suggest that decorations are arranged in such a way as to create an optical illusion, making the male appear larger during performances. Some bowers are even domed. The bowers serve as a stage where males can perform a display and possibly a place where females can visit and be protected from harassment by other males.

The male Jackson's Widowbird creates a display arena in the grassland, where he stomps down a circle of grass from which to perform an energetic bouncing display.

Mating success in Satin Bowerbirds is correlated with bower construction. Males with neat, densely constructed bowers attract and mate with more females. Researchers have found that the number of blue feathers decorating bowers also predicted mating success. Numerous studies on Satin Bowerbirds provide evidence that females are choosing to mate with males based on attributes of bower construction and decorations. Bower quality is associated with age in Satin Bowerbirds, which may be especially important to females in this long-lived species. By mating with males with the best bowers, females are potentially choosing genes that will make their offspring live longer. And, since male bowerbirds do not contribute to parental care, good genes are the only benefit for a female from her choice of mate.

Male Satin Bowerbirds build and decorate bowers with items found in the forest. They will sometimes destroy the bowers of other males, or steal decorations from them.

A male Superb Lyrebird performing his coordinated song and dance routine. The large vocal repertoire of lyrebirds often includes precise mimicry of sounds, both natural and human-made.

During complex dances and other behavioural displays, birds are often using multiple channels of communication. At the most basic level, males with colourful plumage that are visibly perched and singing are displaying multiple ornaments at the same time. But often a displaying male is coordinating the vocal and visual displays.

Researchers have described in detail the coordination of dance moves with vocalisations patterns in Superb Lyrebird males. Males perform solo dances at display sites where females gather and select the mate with the sweetest moves. Males perform complicated dances in which they use movement, plumage and vocalisations. By carefully examining the sequence of events during male dances, researchers were able to determine that males were precisely coordinating dance moves with vocalisations. Males have large vocal repertoires but were only using a few types during dances. In addition, certain types of vocalisations were always associated with one dance move – coordination of separate components of displays in such a manner may involve learning. Song and dance are independently associated with high-quality males in other species because they are traits that are costly. Therefore it is likely that only the highest-quality males are capable of coordination and females are choosing those males that are best at coordinating song and dance.

The ability of male Superb Lyrebirds to coordinate song and dance moves is impressive, but probably one of the most impressive – and mysterious – coordinated displays are the dances of Long-tailed Manakins. Males display on leks but, instead of a solo dance as in other lek displays, males will coordinate with each other to perform as a troupe. Teams of males perform together in the presence of a female. Performing the display along a low branch – the dance perch – males will leapfrog over each other in a fluttering jump while vocalising. The performance troupes usually have two males, an alpha and a beta, but often auxiliary males will join, increasing group size to up to fifteen males! After group performance, the beta male and any auxiliary males leave the area, allowing the alpha male to perform a solo, which eventually leads to copulation for the alpha male. Display rates of troupes are correlated with copulation rates. The troupes that give the most hops, flutters and calls receive the most female attention and get the most copulations for the alpha male.

But what is in it for the beta and auxiliary males? There are several possibilities that might explain such a unique level of coordination among males which should be competing with each other. Cooperating males could be related. Therefore, by helping each other, they are spreading their genes indirectly through the success of the alpha male. Or it could be that sometimes males are the alpha and sometimes they are the beta, and so, by helping in one dance performance, they could themselves be helped in a subsequent dance performance.

Male Long-tailed Manakins have an elongated pair of central tail feathers. Most passerines have 12 tail feathers (with some variation).

to reproduce as beta males. Inheriting a display location is advantageous because females will reliably return to favoured lek sites for mating. The only chance that males have at reproductive success is to become a beta male at a preferred lek site and wait for the demise of the alpha male. Becoming an alpha male can take 10 or more years. So delayed plumage maturation may be an advantage – males do not need to invest in expensive plumage until they are older and perhaps closer to reaching alpha status. During the intervening years, a male can join a troupe as an auxiliary male, eventually becoming a beta male. Alpha and beta males that form longer associations are better at performing together, and females prefer to mate with alpha males that give the most coordinated performances. Long-term strategies that involve delayed plumage maturation and long-lasting associations with alpha males increase the opportunities for young males.

In addition to using exaggerated signals in male–female communication, birds often use the same kinds of signals in an entirely different context: to repel rivals, such as in the territorial and dominance interactions. Similar to the way males try to woo females, and females might benefit from signals that indicate male quality, birds in competition with each other try to intimidate their rivals, and they may benefit by knowing the strength or motivation of rivals before becoming involved in a risky challenge. In the next chapter, we discuss how communication is used to mediate fights.

However, there is no evidence to support either of these hypotheses. Alpha and beta males are not any more related to each other than they are to other individuals in the population, and beta males only become alpha males after the alpha male disappears. It does not appear that beta males are benefiting by being related to the alpha male or that alpha status is reciprocated. The resolution of this unusual scenario is likely related to the extreme life history of male Long-tailed Manakins. They are a long-lived species, which is unusual for passerines, and they take an extraordinarily long time reaching adult plumage – at least 4 years. Most males never get a chance to become the alpha male in performing troupes. Only alpha males reproduce, and only rarely do beta males get a chance

TERRITORIALITY AND DOMINANCE

Bird plumes and songs are some of the most beautiful sights and sounds in nature, and have long been an inspiration to musicians and poets. But while they may be beautiful to human observers, and attractive to potential mates, one of the main functions of such signals is to threaten and intimidate rivals, either to establish dominance or to defend the signaller's territory.

COMPETITION AND TERRITORY DEFENCE

All organisms compete for limited resources, such as food, nesting sites, display sites or mates, and some animals will fight to gain access to and defend their resources.

When the competing individuals occupy roughly the same space, dominance relationships often result, and dominant individuals get priority access to resources. Territoriality is competition between individuals to get exclusive use of a space (the territory). However, fighting is dangerous even for the strongest competitors, and thus, many animals will use signals to communicate their dominance or territory ownership and resolve disputes, rather than resorting to outright combat.

While both visual displays and songs can be used in dominance and territory communication, for many birds, song or other vocal signals are the primary signal used in territory defence. Song can be broadcast over long distances and in dense habitats, such as forests, where visual signals would be less useful. Non-vocal acoustic displays can function in a similar way, such as the drumming signals of woodpeckers. Visual displays in territory defence are most frequently used in cases where the territories are extremely small, such as with Sage Grouse where males display for females in large aggregations called leks. Visual displays may also be used when males are unable to resolve their disputes with long-distance signals and eventually approach each other.

The territory-defence function of song has been recognised for hundreds of years, dating back at least to Gilbert White of Selborne (1789), but it was not directly, experimentally confirmed until we had the ability to record and play back song in the field, a method often referred to as a 'playback experiment' (see 'Playback

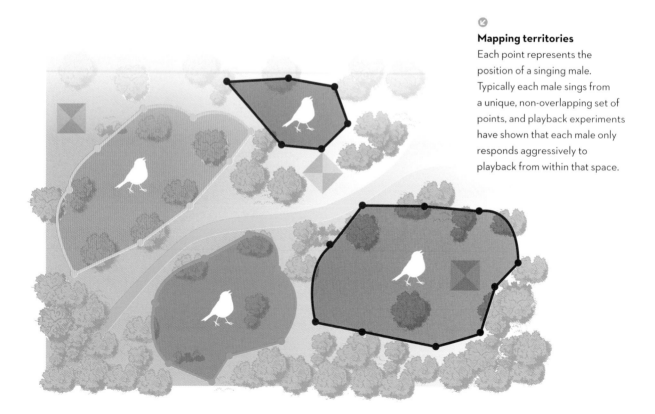

Mapping territories
Each point represents the position of a singing male. Typically each male sings from a unique, non-overlapping set of points, and playback experiments have shown that each male only responds aggressively to playback from within that space.

The Peruvian Warbling Antbird sings to defend its territory against others of its own species, but unusually, it also uses its song in defence against another species, the Yellow-breasted Warbling Antbird.

experiments', page 92). Playing a recorded song to a territorial bird immediately makes it clear that the song of a bird of the same species (a conspecific) is antagonising – the territory owner often, but not always, flies in, searching for the intruder. The territory owner sings and displays, and, when they can't find the phantom intruder, they often engage in a long bout of singing as if to re-proclaim their ownership of that territory. Using playback experiments, we can investigate all sorts of questions about the function of song by assessing a bird's reaction to their own species' song, another species song, a neighbour's song, a stranger's song or different versions of the same song.

As most birdwatchers realise, each species of bird sings a unique song or set of songs. In most cases, birds only use their songs in territorial communication with members of their own species, are only strongly territorial against their own species and will only respond to playback recordings of their own species. There are notable exceptions, particularly among closely related species who share similar habitats in parts of their ranges. For example, some North American warbler species show

interspecific territoriality, defending their territories against both their own species and closely related species with whom they overlap. There are also multiple cases in mountain ranges around the world where a species from a low elevation is territorial against a species from a higher elevation where their ranges meet.

An interesting, recently described case involves South American antbirds, the Peruvian Warbling-Antbird and the Yellow-breasted Warbling-Antbird. These two species are territorial towards one another where their ranges overlap, and the songs of males are more similar where they overlap compared with where they do not. It seems as if the songs of the two species have converged to facilitate territorial communication between them, perhaps because of the benefits of avoiding unnecessary combat.

THE FUNCTION OF SONG
IN TERRITORY DEFENCE

Song is broadcast over long distances, which makes it an effective signal for territory defence, but also raises the question of how song can be such an effective threat. If another bird wants to intrude, the territorial owner is not necessarily in position to present a physical threat. The potential intruder cannot immediately see how big, strong or aggressive the territory owner is. So how does song alone actually stop intruders from entering a singer's territory?

The response of birds to playback of conspecific song makes it immediately clear that territory owners are trying to keep intruders out of their territory, but that doesn't automatically mean that the song is a signal to others to keep out. Perhaps the best way to demonstrate this is a speaker replacement experiment, in which the territory owner is removed from the territory and replaced by speakers playing song. It is then possible to compare the rates of intrusion on the territory where song is being played with the control territories where the male has been removed but no song is being played. Speaker replacement experiments are difficult and time-consuming, not to mention that they are highly disruptive to the lives of the birds who are removed from their territories, and so they haven't been tried too many times.

However, speaker replacement experiments work: an experiment with Red-winged Blackbirds found that territories with the males removed and replaced with speakers playing song had fewer intruders than the control territories. A large repertoire of songs (eight songs in the case of Red-winged Blackbirds) does an even better job of keeping out intruders. This experiment is particularly interesting because this species often lives in dense neighbourhoods of relatively small territories, and they also use a lot of visual signalling, and yet song alone is enough to keep out some intruders.

⬅
Male Red-winged Blackbirds use their plumes and their songs to defend their patch of marsh against other males.

➡
The Bokmakierie is named for its songs, which are produced as male/female duets, and which may function in joint territory defence.

The Yellow-rumped Cacique and Red-breasted Blackbird are members of the American blackbird family, Icteridae, in which larger species tend to have lower frequency vocalisations.

SONG AS THREAT

For Red-winged Blackbirds, their large repertoire of songs appears to be a more effective threat than a small repertoire. If song functions as an effective 'keep out' signal, then the potential intruders must believe that the song is an honest representation of the singer's willingness and ability to fight in defence of the territory. In other words, we would expect that songs give rivals some information about the motivation and quality of the singer.

Body size is likely to be a trait that has a big influence on dominance and fighting ability, and in some cases, it appears that songs could give honest information about the body size of the singer. Exactly why body size and vocalisation frequencies are correlated is not clear, but large body size may mean larger vocal organs, which then allow for songs at lower frequencies to be produced. The correlation between large body size and low frequencies appears to be widespread. In New World blackbirds (Icteridae) and New World doves, comparisons across the groups find that larger species have lower-frequency vocalisations. In White-browed Coucals, both males and females sing, and larger individuals produce songs at lower minimum frequencies. A similar phenomenon was found in a study of Long-billed Hermits, a lek-breeding hummingbird, in which larger males sing songs at lower frequencies.

A similar pattern relating body size and vocalisation frequency has been described in the order Columbiformes, the doves and pigeons (photo shows White-winged Dove).

In the White-browed Coucal (a member of the cuckoo family), both males and females sing, which is a much more common pattern than many people may realise. In this species, the frequency of the vocalisation may provide information on the size of the singing bird.

European Robins show individual variation in how aggressively they fight against territorial intruders. The territory owner typically wins the fight, perhaps because it has more invested and is more willing to fight.

This male Swamp Sparrow in song throws his head back and opens his bill wide, illustrating some of the physical challenges of producing a good song.

Body size could be indicated by other vocal features as well. In Common Nightingales, larger males have larger song repertoires. In Swamp Sparrows, larger males perform more physically difficult, higher-quality songs. In these cases, it is even less clear why body size is related to signal quality. It could be that body size simply indicates overall health or genetic quality, and males in better condition for whatever reason can learn more songs or sing them with more skill.

Non-vocal acoustic signals and visual signals could work in a similar fashion. Male Black Grouse display in large leks, and most of their displays might be thought of as only intended for females, but males fight furiously to establish prime display positions at the centre of the lek. Studies have found that one striking male trait – their bright red eye combs – correlates with body size, and larger males have larger combs. In all such cases, a bird could listen to the song, or assess the visual signals of a rival, and gain valuable information that might influence whether they fight.

Many studies in a wide variety of animals have found that larger individuals tend to have an advantage in fights, but this does not always appear to be the case.

In Long-billed Hermits, larger males are more likely to obtain territories, while in Black Grouse the size of a male's eye combs does not seem to predict his likelihood of winning a fight. Nature is complicated – it might make logical sense for bigger males to have bigger signals and to win more fights, but it is important to remember that just because one can reach a logically valid prediction does not mean that prediction will be supported by the data.

In many cases, territorial and dominance-related signals might be honest, not because of some physiological cost associated with producing the signal, but because of social costs associated with a signal. Territorial song may be an honest signal of willingness and motivation to fight, rather than a signal of fighting ability.

However, fighting is dangerous. From the point of view of the signal receiver – a neighbour or potential intruder – a territorial song indicates that the territory owner is present and ready to fight, and the mere presence might be enough in some cases to convince an intruder to look elsewhere for resources. The signaller – the territory owner – must also be prepared to pay a 'receiver retaliation cost'. The territory owner

⬆
Two male Black Grouse battle for position on a lek, with their bright red eye-combs clearly visible.

⬆
Larger male Long-billed Hermits sing lower-pitched songs, and occupy prime spots at the breeding lek where they display to females.

already has a greater investment in that space, and so should be more willing to fight. Unless there is a huge difference in fighting ability between the territory owner and the intruder, we would expect the territory owner to be more willing to fight, and we would expect the intruder to be quicker to give up. In fact, in many animals that have been studied, such as European Robins, there are strong 'prior residence effects' – individuals already present on their territory tend to show competitive superiority, even if they do not appear to be the larger, stronger individual.

ESCALATION

When a territorial dispute cannot be settled by singing from a distance, many birds go through a series of behaviours that seem to represent increasingly aggressive signalling – this phenomenon is known as escalation. At each step there is a chance that the signals exchanged will do their job, and advertise fighting ability or aggressive intent clearly enough that the opponent will back down, and the dispute will end. If not, the dispute may lead up to an actual physical fight.

The exact sequence in an escalation will differ from one species to another, but the diagram below sets out expected behaviours in the escalation.

Carolina Wrens live together year-round on their territories. Males and females look similar, and both have vocalisations used in territory defence, but the song of the male is a much more familiar sound, ringing out through forests in eastern North America.

Escalation
Song Sparrow disputes start with a song, then may escalate to wing waving (near right), soft song and eventually a physical fight, which can be vicious.

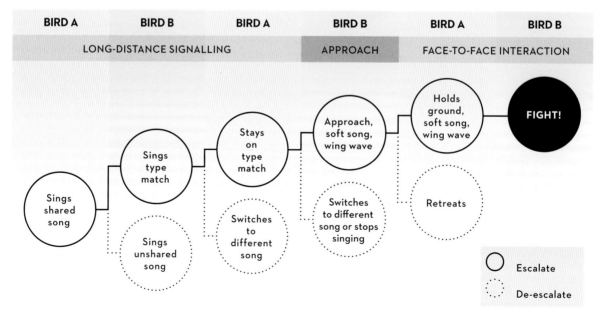

BIRD A	BIRD B	BIRD A	BIRD B	BIRD A	BIRD B
LONG-DISTANCE SIGNALLING			APPROACH	FACE-TO-FACE INTERACTION	

Sings shared song

Sings type match

Sings unshared song

Stays on type match

Switches to different song

Approach, soft song, wing wave

Switches to different song or stops singing

Holds ground, soft song, wing wave

Retreats

FIGHT!

○ Escalate
⸭ De-escalate

COUNTERSINGING

This is often the first step in the process of escalation. Countersinging simply means one territory owner singing in response to another territory owner so that the song bouts of the two birds broadly overlap in time. Countersinging is a common phenomenon that can be observed in species such as Carolina Wrens. A neighbourhood of Carolina Wrens may be quiet, but when one male begins to sing, other males begin to sing in response, and soon a wave of songs spreads through the forest, each male reacting to their neighbours.

SONG OVERLAPPING

Song overlapping is a more precise form of countersinging in which individual songs are produced that temporarily overlap with those of a neighbour. This has been interpreted by many researchers as a signal of male quality, or of aggressive intent, though there is considerable disagreement as to whether this is the case. Song overlapping may seem like a possible aggressive signal because, for human observers, it so closely resembles being interrupted in conversation. If someone continuously and purposefully talks over you, and masks your signal with their own, that tends to make most people angry. However, if two neighbouring male songbirds start singing at an increased rate, the chances of their songs overlapping each will likely increase unless they specifically try to avoid overlapping each other.

MATCHED COUNTERSINGING

In species with song repertoires, matched countersinging may occur as aggressive interactions escalate. When individual males have a repertoire of songs, neighbouring males in the same geographic region may share song types with each other. Imagine that one neighbour has a repertoire of song types (A, B, C, D, E) and another neighbour has his own repertoire (A, B, D, F, G). If bird 1 is singing song type A, bird 2 can match bird 1 by singing a song in 1's repertoire (A, B or D), or bird 2 can avoid matching by singing a song not in 1's repertoire (F or G). Singing a matching song is often associated with slightly escalated encounters, and may represent a way for a bird to address his song towards a specific opponent.

Matched countersinging can be quite noticeable in common garden species with small repertoires, such as Great Tits, Tufted Titmice and Northern Cardinals. Neighbouring males singing at the same time as one another often will be singing the same song type, and it may sound as if a song is echoing throughout the neighbourhood.

If escalation continues, the combatants will approach each other, and may switch to using signals other than their loud territorial song. Postural displays may become particularly important during close-range signalling. Northern Mockingbirds often engage in territorial displays, referred to as boundary dances. In a boundary dance, birds on neighbouring territories face each other and bounce or hop from side to side. At this close range, it is easy to imagine that the two individuals are literally sizing up their opponent, and getting good information about fighting ability, though there are no studies that have specifically addressed this behaviour.

Song Sparrows and Swamp Sparrows engage in territorial displays known as wing-waving, in which highly agitated territory owners will raise one or both wings over their heads and quiver them. In both species, wing-waving is known to be a signal of aggression. Natural fights in the wild can be difficult to observe, and many studies use decoys or taxidermic mounts to elicit aggression. In an experimental setting, males that wing-wave more frequently are more likely to attack a taxidermic mount, and robot Swamp Sparrows built to wing-wave elicit higher levels of aggression.

Northern Mockingbirds are well known for their songs, but they also use plumage-based signals and displays to settle disputes, in which they flash their wings at each other and dance face to face.

These male Northern Cardinals appear to be co-existing peacefully, but males engage in bouts of matched countersinging, where they seem to address their songs to specific neighbours by singing the same song type at each other.

SOFT SONG

If dances and wing-waving don't end the dispute, birds may resort to one of the most aggressive, threatening signals that occurs in nature: soft song. Soft song is often the same acoustic structure as normal, loud territorial song – though it may have a unique structure in some cases – but it is sung at a significantly lower volume, often so quietly that the singer's bill can scarcely be seen moving and the song cannot be heard by a human observer more than a few metres away. In experiments in which the territory owner is presented with a taxidermic mount, soft song has been found to be the signal that most accurately predicts an attack in a wide variety of bird species, including Song Sparrows, Swamp Sparrows, Black-throated Blue Warblers, Brownish-flanked Bush Warblers and Corncrakes.

Singing softly might seem like a surprising way to signal intense aggression. If territorial signals are meant to advertise fighting ability or aggressive intent, how and why should soft song be the ultimate signal? Soft song doesn't seem like it should be harder to produce than loud song, and thus it seems unlikely that soft song provides the most accurate information about body size or condition, or fighting ability. Since the beginning of the twenty-first century or so there have been an increasing number of studies that have documented soft song in a wide number of species, and documented the use of soft song in highly escalated encounters. At the moment, the best evidence suggests that receiver retaliation costs make soft song honest, and yet we still don't understand why it should be soft – it remains quietly mysterious.

FIGHTS

If a dispute cannot be settled with signalling, the last resort may be an actual fight. Some birds are well known for fighting, including Domestic Roosters and lek-breeding birds such as the Ruff, whose Latin name, *Calidris pugnax*, refers to its pugnacious nature. Birds as seemingly delicate as sparrows, warblers and hummingbirds also engage in aggressive combat, and their fights can be quite vicious. In one hummingbird species, the Long-billed Hermit, males have longer, pointier beaks than females, and males use their sharp

Roosters will fight viciously for access to a flock of hens, though the females do not always prefer to mate with the dominant male.

beaks as weapons to jab each other in the neck during fights. If you perform a playback experiment using songs alone, the territory owner flies around searching for an intruder. If you also present a taxidermic mount of a rival, you get to see just how severe a fight can be. The territory owner leaps on top of the mount and bites and pecks at the eyes and the back of the skull of the 'intruder'. An unprotected taxidermic mount used in an experiment quickly loses feathers on the head. In some species of birds, such as Song Sparrows and Swamp Sparrows, adult males are often missing feathers on the tops of their heads, perhaps resulting from a long season of brutal combat.

Playback experiments show that Little Owls recognise neighbours vs. strangers by their calls, and that they know where those neighbours should be – a neighbour in the wrong location is not to be trusted.

NEIGHBOURS AND STRANGERS

Early playback experiments clearly demonstrate that territorial birds use their song as signals in aggressive interactions, but these experiments also revealed that territorial communication includes a surprising *détente*, or cessation of hostilities, between neighbours, even when responses to strangers were aggressive.

Pukeko have complex social lives, and have signals which mediate dominance interactions within the group, as well as signals used as they defend their shared territory against other groups.

During the spring, male Sooty Grouse 'hoot' to attract females, but they also aggressively keep other males away from their territories, and recognise known neighbours vs. potentially intruding strangers on the basis of their hoots.

A study of territoriality in Ovenbirds compared how birds responded to the songs of neighbours on adjacent territories and the songs of strangers on non-adjacent territories, and found that territory owners responded more aggressively to the songs of strangers. Although all Ovenbird songs sound similar enough for a bird or human observer to recognise them, there is enough variation between songs for the Ovenbirds to pick out individual differences. A territorial male Ovenbird can tell the difference between the songs of neighbours and strangers and so can recognise neighbours and strangers by their songs alone. The same signals that mediate fights between neighbours allow for neighbours to recognise each other and reduce aggression.

Discrimination between neighbours and strangers and decreased aggression towards neighbours – often called the 'dear enemy effect' – is very common, and has been described in many species of songbirds, as well as a wide diversity of non-songbirds, including Sooty Grouse, Audubon's Shearwater, Pukeko (New Zealand Purple Swamphen), Little Owl and Alder Flycatcher. It is particularly interesting to note neighbour–stranger discrimination in the non-songbirds because the birds are thought to have innate vocalisations – they do not learn the vocalisations they produce, yet apparently they can learn to recognise the vocalisations of others.

TERRITORIALITY AND DOMINANCE

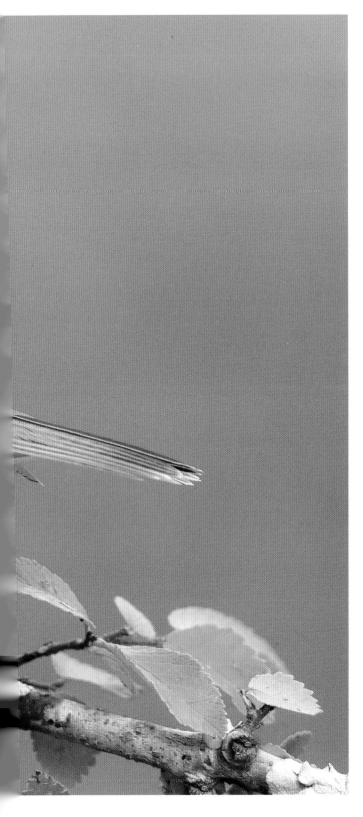

Some birds have been shown to be capable of an even finer level of recognition, in that they can recognise individual neighbours, as opposed to just differentiating between neighbours and strangers. Experiments demonstrating individual neighbour recognition can be undertaken in several ways, but generally involve performing playback experiments of an individual neighbour's songs both at the territory boundary shared with that neighbour (a correct boundary) and at a different territory boundary (an incorrect boundary). If the territory owner responds more strongly to a neighbour at the incorrect boundary than to the neighbour at the correct boundary, then we have evidence that the territory owner doesn't simply recognise neighbours and strangers, but recognises where each individual neighbour should be.

Hooded Warblers provide one of the most spectacular examples of individual neighbour recognition. Experiments have shown that Hooded Warblers can recognise individual neighbours within a breeding season. But those male Hooded Warblers leave their breeding grounds at the end of the summer and migrate to Central America for the winter – when they return the following spring, they still retain the memory of the songs of their neighbours from the year before, and respond more strongly to neighbours who are in the wrong spot.

This male Hooded Warbler may remember his neighbours from one year to the next, allowing him to quickly transition from fighting over territory boundaries to finding a mate and raising offspring.

THREAT ASSESSMENT

But why should a territory owner respond more strongly to a stranger than a neighbour, or more strongly to a neighbour in the wrong place? The explanation for these results could be that territory owners simply habituate or reduce their physiological responsiveness over time to the songs they hear most often. But rather than habituate to all songs, the birds specifically habituate to the songs of neighbours.

There are two main evolutionary hypotheses proposed that explain reduced aggression towards neighbours. The first – the relative threat hypothesis – is based on the relative threat posed by neighbours and strangers, and proposes that the neighbours a bird hears over and over again are less of a threat to its territory than a stranger. After all, a known neighbour, despite being the closest rival, already owns a territory, while a stranger may not and so may be interested in stealing another bird's territory.

The second hypothesis – the relative familiarity hypothesis – proposes that the neighbours are more familiar to the territory owner and it has interacted, and perhaps fought, with them many times. A territory owner is not going to get into an escalated fight with a known neighbour because they've already had that fight before,

and they know the likely outcome of another fight with a neighbour, while they have not established a relationship with a stranger.

The common theme to both hypotheses is that recognising neighbours allows territorial birds to save time and energy by avoiding unnecessary territorial fights. Instead of fighting with everyone, males use their recognition abilities to direct their aggression where most appropriate, and so can spend more time and energy on other important activities such as foraging or attracting a mate.

The relative threat hypothesis for reduced aggression towards neighbours has generally received more support, in large part from the many cases in which a bird's neighbour, despite being a very familiar individual, may represent an equal or greater threat

⊙

Despite having individually distinctive songs, Spotted Antbirds do not appear to respond differently to the songs of neighbours vs. strangers.

These Northern Gannets illustrate the aggression that can occur between neighbours in a tightly packed colony.

In the high-pressure environment of a Greater Prairie Chicken lek, males show equal aggression towards familiar and unfamiliar rivals.

Buff-banded Rails do not discriminate between neighbours and strangers, perhaps due to a high level of territorial instability in this species.

than a stranger, and they elicit as much or more aggression. For example, familiar neighbours might be as much of a threat as a stranger when neighbours are in close proximity and there is intense competition over resources, such as in breeding colonies or communal display areas, known as leks. In a dense colony of Gannets, neighbours are constantly squabbling over nest site boundaries and stealing nesting material from each other. It's easy to understand why a Gannet might be as aggressive towards familiar neighbours as strangers. Male Greater Prairie Chickens show significant individual variation in the calls that they use during courtship, but males do not respond differently to the calls of neighbours and strangers. It seems that competition for prime spots on the lek may be too intense for them to care which bird is a neighbour and which is a stranger.

Even in territorial systems, the dear enemy effect is not universal. Some species, such as Spotted Antbirds and Buff-banded Rails, show no difference in response to songs of neighbours and strangers. Other species show the opposite of a dear enemy effect – the nasty neighbour effect, a stronger response to neighbours than strangers. A study of neighbour–stranger discrimination in female New Zealand Bellbirds, a species in which females sing regularly, found that females responded more strongly to the songs of their female neighbours, indicating that, in this species, female neighbours represent a greater threat than female strangers.

The relative threat of different neighbours may vary as well. In some cases, the difference in threat could be sex-specific. Male and female Bay Wrens both sing, and they can sing solo, or produce tightly coordinated duets. Female Bay Wrens respond more aggressively to the songs of females than of males, perhaps because another female represents a territorial threat, whereas a male does not. In other cases, even individual neighbours may vary in their level of threat. Song Sparrows respond more strongly to aggressive neighbours than non-aggressive neighbours. Red-winged Blackbirds respond more strongly to neighbours who have previously engaged in extra-pair sexual behaviour with their mate. It is clear that not all neighbours are regarded equally by some birds.

In addition, the relative threat of familiar neighbours can change. All experiments demonstrating individual neighbour recognition show that a familiar neighbour in the wrong location becomes perceived as a stranger – they may be looking to change or expand their territory boundaries and are now untrustworthy. The relative threat of neighbours can also change depending on the timing of the reproductive cycle. Female Song Sparrows are at their most fertile while they are building nests and laying eggs. Like many other birds, Song Sparrows are somewhat promiscuous, and, on average, around 20% of chicks on a territory were sired by a bird other than the territorial male, and most frequently the extra-pair sire is a neighbour. During the fertile period, a territorial male should be at his most vigilant due to the possibility of losing paternity at home, and studies have shown that male Song Sparrows show the dear enemy effect before and after their mate's fertile period, but not during the fertile period when even a familiar neighbour cannot be trusted. Song Sparrows also appear to use their individual recognition abilities to direct aggression towards their most threatening neighbours – either those that are particularly aggressive, or those that have previously intruded on their territories. When viewed in the light of the relative threat hypothesis, the dear enemy effect begins to look like a form of cooperation. Both neighbours benefit from reducing aggression towards neighbours, but the reduced aggression is only mutually beneficial if both neighbours continue to behave in a trustworthy fashion. Any intrusion or act of aggression can cause the dear enemy effect to break down and lead to increased fighting.

PLAYBACK EXPERIMENTS

Playback experiments are a key technique that allows researchers to investigate how song is used in territory defence. Most of these experiments require researchers to record examples of song from multiple sources, which can then be used to simulate the presence of neighbours and strangers, or local birds and foreign birds, or any other comparison the researcher wants to make. Many playback experiments also require careful mapping of the territories of a neighbourhood of birds, so that a researcher can simulate songs coming from specific locations, such as a rival at the boundary of a territory versus a rival intruding into the centre of a territory, or a neighbour in the 'correct' location versus a neighbour in an 'incorrect' location. Over the course of a breeding season, territory boundaries may shift, and some males may even switch territories, but in many species territory locations tend to be stable, even from year to year.

In the Bay Wren, both males and females sing. Females appear to defend the shared territory against other females, while males defend the territory against other males.

APPEARANCE AND DOMINANCE

Visual signals, including plumage-based signals, are also very important in territoriality and dominance in many species. The significance of plumage-based signals in territorial behaviour may be reflected in the correlation between plumage colour and territory quality seen in many birds.

For example, in Northern Cardinals, redder males obtain better breeding territories, with more dense vegetation. In King Penguins, both sexes have patches of yellow feathers on the side of their heads. Within a large breeding colony, pairs with larger yellow patches tend to settle in the centre of the colony, which is likely safer from predators than peripheral areas. American Redstarts are territorial in both the breeding season and on their wintering grounds in the Caribbean. Males with winter territories in high-quality black mangrove habitat have brighter orange tail feathers than males in poor-quality scrub habitat.

Several studies have shown that variation in plumage colour also seems to relate to a bird's ability to compete for territories. For example, King Penguins with

Both male and female King Penguins have brightly coloured patches of feathers that may signal competitive ability, as birds with bigger patches engage in more fights and occupy better, more central locations in the dense colony.

↑
Male Lark Buntings with more black feathers on the body tend to win fights, but males with larger wing patches experience fewer territorial intrusions.

↗
Male American Redstarts with brighter orange tail feathers are found in higher-quality winter habitats, and spending the winter in better habitats may allow them to migrate to the breeding grounds earlier, attract more mates and achieve higher reproductive success.

larger yellow patches tend to show higher rates of aggressive behaviour. However, the King Penguins with larger yellow patches are the ones near the centre of the colony, so they may simply find themselves in more fights than individuals in peripheral areas. Male Lark Buntings have striking black and white plumage, and in territorial fights observed in the wild, males with more black in their plumage tend to be the dominant male in the interaction.

Experimental studies provide even better evidence of plumage signalling fighting ability. A study of Eastern Bluebirds examined the relationship between the males' blue ultraviolet (UV) colour and their ability to attain a nest box. Bluebirds are highly dependent on human-made nest boxes, which allows researchers to manipulate nest site availability and territory quality. A set of nest boxes was put out in the early spring and, after those nest boxes were claimed, a second set of nest boxes were added to the study site. The males who obtained the first set of nest boxes had more colourful UV plumage than the males who got the later nest boxes. This suggests that the more colourful males had their first pick of territories, and the less colourful males had to wait.

↶
Eastern Bluebirds compete intensely for nest boxes. Birds with more colourful ultraviolet plumage, invisible to the human eye, win more nest sites and fledge more offspring.

Male Collared Flycatchers have a white forehead patch which is displayed in territorial fights. A study of Collared Flycatchers (which also use nest boxes) captured territorial males early in the breeding season, and held them in captivity for a few hours. While they were gone, other males came in to try to take over the now-vacant territories. After a new bird arrived, the original owner was released, and chaos ensued. Males that regained their territory had larger forehead patches than the males that failed.

However, it should be noted that other studies have found that plumage signals may relate to territory quality, and yet are not related to aggression. One such example was found in a study of Golden-winged Warblers. Male Golden-winged Warblers have colourful yellow crowns, and while birds with yellower crowns held higher-quality territories, they were less aggressive in response to song playback experiments. One possible explanation is that, in some cases, birds with high-quality plumage signals may not have to behave aggressively to win a fight. Perhaps the intruder tends to give up as soon as they see the big, colourful signal.

A series of studies of Red-collared Widowbirds might offer a fuller picture of how plumage signals can function in territory defence. Male Red-collared Widowbirds compete to gain possession of territories where the females will nest, though the males provide no parental care. Some males fail to obtain a territory, and spend the breeding season as 'floaters', hoping to find a vacancy. Territorial males have larger, redder collars than floaters, which suggests that the red collar is a signal of fighting ability.

Experimental manipulations of colour support this idea. Males that were painted to have larger red collars claimed larger territories than those whose collar size was reduced or colour dulled. The males with the painted large red collars also experienced fewer intrusions on their territories, and spent less time fighting than males with smaller collars. An experiment presented territorial males with taxidermic models of male widowbirds and found that males with larger, redder collars responded more strongly to the models, but also, territory owners tended to be less aggressive

↑

The bright crown of this Golden-winged Warbler seems like an obvious signal, but the intended receiver of the signal and the message it carries is not entirely clear.

→

The Red-collared Widowbird shows off multiple signals for different audiences: the red collar is involved in male–male communication, and the long tail in male–female communication.

towards models with larger, redder collars. The conclusion seems to be that males with bigger, brighter collars can acquire better territories, but spend less time fighting because rival males see the red collar as a reliable, honest signal of fighting ability and steer clear of the toughest males. But there's one other twist: female Red-collared Widowbirds don't care about the red collars of males – they prefer the males with the longest tails. The red collars appear to function only in male–male competition.

Bold colours may act as signals outside of the breeding season as well, such as the black bib on this Harris's Sparrow.

White-eared Hummingbirds with bigger 'ear' stripes may be dominant over individuals with smaller stripes, despite not being larger in body size.

Black-crested Titmice raise their crests during agonistic interactions. During the summer, when population density is high and competition is fierce, males with longer crests had greater access to feeders.

Outside a territorial context, many birds have plumage signals that may be particularly important in dominance interactions. These plumage signals are often referred to as 'badges of status'. For example, the melanin-based, black bib of a Harris's Sparrow varies considerably in size and colour between individuals. Individuals with larger badges tend to be behaviourally dominant over individuals with smaller badges, particularly in winter flocks. Being able to signal dominance, and recognise dominant and subordinate individuals can be very beneficial as it allows group members to avoid fights they will obviously lose, and so all individuals can spend less time on fighting, and more time on self-maintenance activities such as feeding.

Many examples of badges of status have been described over the years, such as the red bill shield of a Pukeko, the length of the crest of a Black-crested Titmouse, the supercilium stripe in White-eared Hummingbirds and, as we have seen, the red patches on Red-collared Widowbirds. In all cases, the size or colour of the patch has been found to correlate with

behavioural dominance or access to food. But why do badges of status work and why do other birds behave subordinately to individuals with larger badges? An obvious hypothesis would be that larger or darker-coloured patches correlate in some way with the fighting ability of the individual, and sometimes researchers have found this to be the case. However, the honesty of badges of status, particularly melanin-based black badges, has been a contentious issue among researchers. It is not clear if the badge correlates with dominance because badge production is in some way costly, and thus only high-quality individuals or those in good condition can produce the badge. It could also be the case that large badges tend to invite aggressive behaviour, and so only high-quality individuals can afford to wear the badge.

Research on badges of status has resulted in occasionally confusing results. Studies of a range of species that have presented birds with models that varied in badge size sometimes find that models with larger badges are avoided, as if the models are dominant and threatening, but in other cases the models with big badges tend to elicit more aggression, which could also be a reasonable response to a dominant and threatening rival.

Other studies experimentally manipulated the sizes of badges of live birds to examine the function of these signals. In some cases, dyeing a bird's feathers to give them a larger badge caused them to rise in dominance status. This would only make sense if other birds accepted the badge as an honest signal and didn't challenge the bird with the fake high-status badge. If the fake badge elicited more aggression, the fake would soon be found out.

So, if a big badge is a signal of high status, should individuals with high status elicit high aggression, or should they be avoided? This may simply be an area in which more research on more species is needed before we can understand the overall pattern. There may be important differences between responses of high-status and low-status individuals, or familiar and unfamiliar individuals, to the badges of others.

TERRITORIALITY AND DOMINANCE

In Golden-crowned Sparrows, individuals with larger gold or black plumage patches win contests among strangers. Being among familiar flock mates has its benefits – less time spent fighting and increased foraging success.

Both male and female Pukeko have conspicuous red frontal shield ornaments which they display during aggressive interactions.

For example, studies in Golden-crowned Sparrows have found that the black and gold feather patches on the head act as badges of status, and individuals with larger patches tend to be dominant over those with smaller patches. Even in birds with experimentally altered colour patches, birds with artificially large patches tend to be dominant over those with small colour patches. However, these fake badges only work in cases where the two birds don't know each other. Familiar flock mates seem to recognise each other, and an artificial change in their badge doesn't change the dominance relationship they've already established. But for birds that are strangers, the badges – even fake ones – appear to be important signals of dominance. There may also be important differences between the use of such signals in territorial contexts and in winter flocks. What is true about the signalling system for one species may not be true for another, and we may have not yet identified the key variable that explains the difference.

PUKEKO'S BADGE OF STATUS

The Pukeko, the New Zealand Purple Swamphen, lives in complex social groups that involve a combination of competitive and cooperative behaviours. Individuals compete against each other for mating opportunities and form dominance hierarchies, but also cooperate in raising chicks and defending the group territory. The bright red frontal shield of the bill appears to be a badge of status. Individuals with larger red shields are typically the most dominant members of the group.

An interesting thing about the red shield is that, unlike feather colour or patch size, its size can change quickly. Experiments that reduced the size of a bird's shield by painting the edges black to match the head feathers immediately caused the birds to be attacked by group mates more often, and they lost rank. The feedback between lower rank and signal size was quick. A week later, the manipulated birds were caught and measured. Amazingly, the loss of status, and perhaps the stress associated with being attacked more often, caused the manipulated birds to produce a smaller shield. They were now signalling a lower status.

PARENT–OFFSPRING COMMUNICATION

The songs and displays of birds enable them to gain territories and find mates. Finding a mate may require intense competition against other suitors to convince a sceptical partner that a bird is the best-quality mate available. But a bird that is successful in these stages can then settle into bringing up young. As the common interest of both parents and their offspring is for those chicks to thrive, family life can involve a lot of straightforward communication, cooperation and altruism. However, it can also involve a considerable amount of conflict, competition and even deception.

CONFLICT IN BIRD FAMILY LIFE

The conflict in bird family life comes from several sources. First, parents and offspring are often in conflict. Many young birds require a tremendous amount of parental care, and providing parental care can be exhausting.

While the young birds want to get as much parental care as they possibly can, parents are under selective pressure to not tire themselves out taking care of their current brood if it takes away from their ability to have a future brood. Second, siblings in the nest are potentially in conflict, as each chick may care more about its own survival than its siblings', and want to gain more than its fair share of food. Conflict between siblings can be particularly intense if food is in short supply, and there isn't enough for all the young birds in the nest to survive, or if the chicks in the nest are not all related to each other to the same degree. Extra-pair copulations (EPCs)

mean that some chicks in the nest might be siblings, but others might only be half siblings. Intense competition can potentially result in one chick being killed, either via parental neglect, a parent actively killing off a chick or siblicide, in which one chick kills another. When chicks are in competition, parents may be faced with the decision to feed all the chicks they have, feed those that are most in need or feed the most valuable.

This European Greenfinch is facing a dilemma experienced by many bird parents: which of these mouths should I feed?

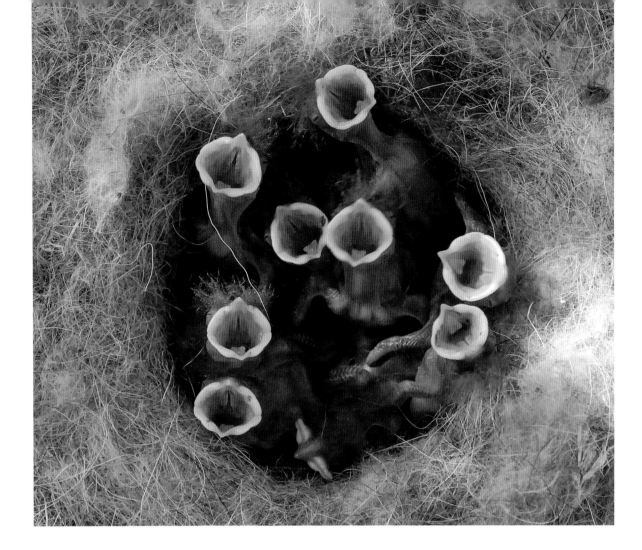

BEGGING SIGNALS

Parents and offspring have forms of communication that have evolved in response to these situations. The primary signal, known as 'begging', influences the amount of parental care that each chick receives. Begging behaviour includes the sounds, postures and colours of baby birds trying to attract the attention of their parents. Begging might not initially seem like a signal on par with songs, plumes and dances, but a closer examination suggests that begging is not just incidentally noisy and noticeable, it has evolved to be especially noisy and noticeable. Begging calls are not merely sounds of exertion, they are attempts to get attention.

Young songbirds begging to be fed. The colour and enlarged flanges of the gape, as well as the extended necks and begging calls, are all attempts to communicate to the parents and influence which chick gets fed.

Equally, the bright red mouth and fleshy orange gape of many baby songbirds is not a simple, physiological necessity, it is an attempt to be seen. Even the exaggerated way that baby birds lift their heads and quiver has been shaped by selection to be an effective signal. Begging has attracted the attention of biologists for decades because it has the potential to be an extremely costly signal – it can be so noisy and noticeable that it attracts predators.

None of these things would be necessary if parent birds calmly went around the nest making sure every chick received food. Instead, almost immediately after the chicks hatch, they are under pressure to signal to their parents for further parental investment, and the parents are faced with decisions to make about who to feed. The recognition that parents and chicks, as well as siblings in the nest, are in conflict and have divergent interests about which ones get parental care has led to a huge number of studies on begging behaviour, asking questions such as: What does begging signal – the need of a chick, or the quality of the chick? Do parents want all their chicks to survive, or their best chicks to survive?

NEED

The hypothesis that begging signals offspring need is perhaps the most straightforward idea. A hungrier chick should signal more, and a satiated chick should signal less. This hypothesis can be experimentally tested by depriving chicks of food and looking for changes in begging behaviour. In Rock Doves, Yellow-headed Blackbirds and others, chicks that are experimentally deprived of food spend more time begging than non-deprived chicks, and, in return, are fed more often.

In Tree Swallows, food-deprived nestlings produce longer begging calls at a greater rate, and playback experiments revealed that parents respond more strongly to the sound of a food-deprived chick. In Canaries, Greenfinches and Bullfinches, the colour of a chick's mouth becomes brighter when it is food-deprived. In Crested Auklets, Parakeet Auklets and Horned Puffins, food-deprived chicks produce calls that are higher in frequency than those of non-deprived chicks.

This Tree Swallow chick has met a parent at the door, perhaps signalling its hunger through its begging behaviour.

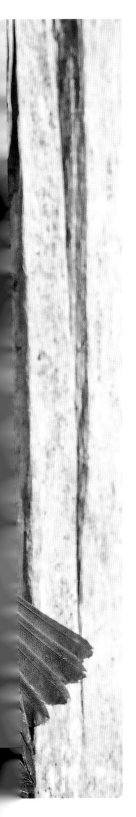

In Eurasian Bullfinches, the colour of a begging chick's mouth may signal its hunger, though the position of the chicks in the nest and how outstretched their necks are may have a big effect on who gets fed.

Crested Auklets (alongside their Least Auklet neighbours) breed in crevices on sea cliffs. The Crested Auklet chick produces higher-pitched calls when it is hungry, though there is only a single chick in the nest, so the hungry chick is merely communicating its need, not trying to outcompete any siblings.

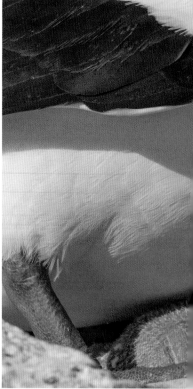

Interestingly, Auklets and Puffins have only one chick per nest, so in these cases begging should only indicate need, as the chick is not in competition with other siblings for parental attention. The chick may still seek more parental investment than the parents are willing to give; the parents have future reproductive opportunities and future nestlings to consider.

Some species of birds set up a 'natural experiment' in food deprivation, by laying one egg per day, but starting to incubate before the last egg is laid. As a result, the chicks hatch on different days – a phenomenon known as hatching asynchrony – and the last hatched chick starts life younger and smaller than the rest of its nest mates. In Barn Swallows, the youngest chick is typically the smallest, and spends more time begging than the older chicks. Studies have provided strong evidence that begging behaviour can signal the hunger or need of the chicks.

↑
Barn Swallows' parental care may continue for more than two weeks after the chicks leave the nest.

↗
This Masked Booby chick has a head start over its yet unhatched sibling in the egg behind it. This advantage will likely lead to the older chick being the only survivor from the brood.

→
Even in successful Eurasian Hoopoe nests, often the youngest, smallest chick will die before fledging.

QUALITY

At the same time, parents don't always care about offspring need, and this can be particularly clear in species with hatching asynchrony, as well as species that typically show brood reduction, where not all chicks that hatch survive to leave the nest and fledge. The most extreme cases are in siblicidal species, such as the Masked Booby, which typically lays two eggs per brood, though only one chick survives to independence as the oldest chick relentlessly attacks its younger sibling.

Similarly, in Blue-footed Boobies, the largest chick begs the most, is fed the most and the smaller, hungrier chick is left to die. In Hoopoes, another species with hatching asynchrony, the parents appear to preferentially feed the largest chick that begs, and only feed the smaller chicks when the largest chicks ignore parental attempts to feed them. In the Yellow-headed Blackbird mentioned above (also an asynchronous-hatching species), although hungry chicks beg more and are fed more, parents also pay attention to the size of their chicks, and small chicks, even when hungry, are never fed as much as the larger, older chicks.

Rufous-tailed Scrub Robins don't show the level of hatching asynchrony and brood reduction seen in the above examples, and yet slight asymmetries in size between the chicks are exacerbated by the fact that larger chicks beg more intensely, stretch their necks up higher and get fed more often than the smaller chicks. These sorts of studies provide evidence that begging behaviour can signal the quality of the chicks, and that, in some cases, parents may bias their feeding in favour of the larger, healthier chicks.

Perhaps the clearest cases in which chicks are signalling their quality are found in species where the young birds don't just make noise, but also have signals such as bright plumage, mouth parts or skin. The conspicuous features of some chicks have led researchers to suspect that they are ornamented for signalling in some way, just as we would expect for the brightly coloured anatomy of an adult bird.

A Barn Swallow parent approaching the nest is immediately faced with an onslaught of signals from chicks, with each one trying to get more than its fair share of food.

The yellowish bills of these Little Owl chicks may be telling the parents which chick is the healthiest and who should get fed.

For chick signals, the most likely receivers are the parents. Barn Swallow chicks have bright, colourful gapes but rather than signalling hunger, as in the cases of the finches mentioned above, it appears that brighter gapes are a signal of a healthier chick, and the healthier chicks are fed more than the others. Little Owl chicks have noticeably yellow bills, and chicks with yellower bills tend to be larger. Experimental manipulation of bill colour revealed that, in a large brood of chicks, parents prefer to feed the chicks with yellower bills. In European Coots and American Coots, chicks sport bright red and orange bald heads, and a ruff of waxy, orange feathers around their necks. Researchers manipulated chick ornamentation by clipping off the orange feathers of some chicks, making them appear duller. Parents responded by feeding the ornamented chicks more than the experimentally non-ornamented chicks, and the ornamented chicks grew faster and had higher survival rates.

A wide comparison of coots and their relatives, the rails, in the family Rallidae, found that in most rails, and particularly in the most primitive species, the typical colour of chick plumage was plain black. Chicks tended to be more ornamented in species with larger brood sizes and in polygamous species, where chicks in the nest may not all be full siblings, so that some chicks are less closely related than others in the same nest. In a larger brood, or a less related brood, the intensity of competition between chicks is expected to be greater, resulting in stronger selection for ornamentation of chicks to show their quality.

It seems clear that begging can be a signal of need in many cases, and in others a signal of quality. But why would this be? Is there any pattern to which chicks beg owing to need, and which ones beg to show off? Luckily, begging has been studied so frequently, and in such a wide diversity of birds, that we may now understand the range of begging strategies. A study comparing begging behaviour among 143 species of birds concluded that for species in predictable, resource-rich environments, chicks in worse condition beg more – begging is a signal of need in good habitats. Parents respond by feeding the chicks in need, and so parents can increase their reproductive success by increasing the number of young that fledge. For species in unpredictable, poor environments, vocal begging is less important and so parents pay less attention to begging.

The diversity of colouration of rail chicks ranges from plain black in the Water Rail (top left) to a colourful bill in the Purple Gallinule (opposite), a colourful bill and bald head in the Common Moorhen (top right) and a colourful bill, head and orange plumes in a Eurasian Coot (above). This diversity might be explained by selection on the chicks to show off their quality to their parents.

Instead, parents rely on physical signals of quality, such as the Little Owl bill colour or coot feather ornaments. Parents respond by feeding the chicks that signal they are the highest quality, and so, again, the parents can increase their reproductive success by favouring the chicks most likely to survive and thrive.

Even within a species, parental response to chick signals can differ, depending on habitat quality. In cases in which chicks signal their quality, the effect of their signals may only be important in the worst conditions. In the case of the Little Owl mentioned above, chicks with yellower bills get fed more, but only in large broods, where competition for food is likely to be greater. In small broods, chicks get fed equally, regardless of bill colour. It appears that parents respond to the signal of quality only when they are faced with the prospect of not being able to adequately feed every chick.

RECOGNITION

Begging behaviour might provide a signal that allows parents to feed the right chick – either the hungriest of the brood, or the highest quality of the brood. Sometimes, feeding the right chick means something more fundamental – feeding the chick that you are related to. You don't want to feed the hungriest chick if it's not your chick. For many birds, being able to recognise their own chicks is not a critical skill. If the parents know where their nest is, there is little chance that a parent will confuse other chicks for their own. The need for parents to recognise chicks, or for chicks to recognise parents, and the evolution of signals to allow recognition occurs in a few specific situations in which the chicks at or near the nest are not necessarily related to the parents – when brood parasites are present, and in large, crowded breeding colonies.

BROOD PARASITES

Brood parasites, such as cuckoos and cowbirds, are birds which build no nest of their own, and instead lay their eggs in the nest of other species and let the other species provide parental care. When brood parasites are present, some of the chicks in the host nest could be the hosts' offspring, and some could be imposters put into the nest by a parasite. Host parents face the risk of investing their parental care in an impostor, and potentially losing all reproductive success. Despite the obvious benefits of chick recognition, sometimes parents spectacularly fail to recognise their chicks, resulting in absurd sights like a 12 gram Reed Warbler feeding a Common Cuckoo chick ten times its own size. Any signal, visual or acoustic, that allows the parents to recognise their chicks and avoid being parasitised would be strongly favoured by selection.

Superb Fairywrens are frequently parasitised by Horsfield's Bronze Cuckoo. Cuckoo parasitism results in fairywrens raising chicks that are not their own, which puts strong evolutionary pressure on the fairywrens to avoid parasitism, or recognise when they've been

parasitised, and cut their losses when it comes to parental investment. The first step a fairywren could take to avoid parasitism might be to avoid having the cuckoos spot the nest, but, in a multi-year study, anywhere from 19 to 37% of fairywren nests were parasitised each breeding season. Having failed the first test, the fairywrens could abandon their nests when they had been parasitised, and the same study found that females abandon almost 40% of nests that have a cuckoo chick, but virtually never abandon nests that contain only their own young. So how do the females recognise they've been parasitised? This requires at least some form of chick recognition, but doesn't

A Eurasian Reed Warbler feeding a Common Cuckoo chick is wasting a tremendous amount of parental investment.

Cowbirds are brood parasites that are common throughout North and South America. This cowbird chick is ready to receive some parental care from an unsuspecting host.

⬆
Horsfield's Bronze
Cuckoo parasitises
fairywrens and thornbills
throughout Australia.

⬆
Superb Fairywrens have been extensively studied,
due to their fascinating social lives (they are cooperative
breeders), promiscuous sex lives and complex
adaptations for dealing with cuckoo parasitism.

necessarily require a signal. The cuckoo chick typically hatches earlier than the fairywren chicks, and often the cuckoo chick will eject the fairywren chicks, which results in a nest containing just one chick. Female fairywrens will often abandon nests containing a single chick, suggesting that they perceive the single chick as evidence that they've been parasitised.

The begging call of a fairywren chick could also provide a signal that allows parents to recognise their young, but only if the call is distinct from the call of the parasite. Horsfield's Bronze Cuckoo chicks appear to be able to learn at least some aspects of the fairywren chick call. The begging calls of the fairywren and the cuckoo are quite similar. In a cross-fostering experiment, cuckoo eggs laid in fairywren nests were placed in the nests of Buff-rumped Thornbills. When the cuckoo eggs hatched, the cuckoo chicks initially produced calls that were like fairywren begging calls, but very quickly modified their begging calls to sound like thornbill begging calls. The cuckoo chicks, when just a few days old, can clearly learn to change their calls to match their host.

But Superb Fairywrens also have a unique vocal signal, learned by chicks before they hatch, which the parents use to recognise their chicks. Female fairywrens call to their eggs, and the unhatched chicks learn part of her call. When they hatch, the chicks produce this call – a 'vocal password' – which shows the chicks belong to the parent, whereas cuckoo chicks do not learn this call. This signal allows the fairywren parents to discriminate between their chicks and the brood parasite, and make sure the right mouths are fed. It seems that, for the moment at least, cuckoo chicks cannot learn this vocal password.

Visual signals might also help parents avoid brood parasites. An Australian warbler, the Large-billed Gerygone, is parasitised by the Little Bronze Cuckoo. Gerygones can recognise cuckoo chicks, and will actively drag the baby cuckoo out of the nest. The recognition signal appears to be the odd white plumes found on the otherwise naked chicks. The cuckoo chicks are excellent mimics, having the same dark skin colour and white plumes of the gerygone chick, but the gerygone chicks have more feathers. Gerygone parents rarely reject their own chicks, but will reject cuckoo chicks 69% of the time. If the feathers of the cuckoo chicks are experimentally trimmed, the rejection rate goes up to 89%. Different subspecies of Little Bronze Cuckoo mimic different host species, which suggests that host parent rejection has led to the evolution of close mimicry by the cuckoo chicks. While we don't know for sure that the gerygone chick's plumes have evolved to be a recognition signal, it seems possible that the pressure from cuckoo parasitism has led to the elaboration of this trait, and parental rejection may lead to further perfection of the cuckoo chick mimicry.

Another possible visual signal used in chick recognition is the brightly coloured gape found in many young birds. As we have seen, the gapes of chicks appear to act as signals of need in some cases, and signals of quality in others, the brighter mouth being fed. If the gape of the parasite provides a stronger signal of hunger or quality, then the parasite gets fed more. If the brood parasite gape doesn't match the gape of the host species, then the parents might think the parasite chick isn't hungry or isn't worthy. However, chicks of some brood parasites appear to mimic the gapes of their hosts, allowing the parasite chicks to escape detection and get fed as well. In an evolutionary arms race, the host species might develop even more exaggerated gapes, which may allow parents to recognise their offspring, but it may also just be an adaptation to allow the host chicks to outcompete the parasites. In either scenario, the outrageous ornamentation of the mouths of some baby birds could represent a stage in a long-running battle between some species and their brood parasites.

The Little Bronze Cuckoo shown here and the Large-billed Gerygone are locked in an evolutionary arms race, as the gerygone evolves to avoid parasitism and the cuckoo evolves to evade the gerygone defences.

The family Estrildidae is a colourful group of finches found in Africa, Asia and Australasia. The African species, particularly the Common Waxbill, are frequently parasitised by indigobirds and whydas in the genus *Vidua*. A large study of waxbills found that parasitised species tend to have brighter and more colourful mouths than non-parasitised species, and that there is more variation in gape colour of the parasitised species. This pattern is exactly what you'd expect if pressure from parasites is driving the evolution of chick mouth colouration to outcompete the parasites.

The brood parasitic Pin-tailed Whydah lays its eggs in the nest of the Common Waxbill. The whydah chick, top left, has gape markings that closely mimic the markings of the host bird, the Common Waxbill, top right. An adult Common Waxbill is shown on the right.

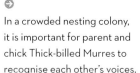

In the darkness of a tree cavity, the bright white mouth flanges of a Northern Flicker chick might help to catch the parents' attention.

In a crowded nesting colony, it is important for parent and chick Thick-billed Murres to recognise each other's voices.

Another possibility is that bright mouth colouration and ornamentation in some species may not be a signal of quality, a signal of need or a signal used for recognition, but instead they may be a signal used for detection. The bright edges of the mouth, called the flange, in particular might help parents find the mouths of their chicks in the dark. Studies in cavity-nesting birds such as the Northern Flicker, Pied Flycatcher, Great Tit and House Sparrow have shown that chicks with experimentally darkened flanges were fed less and gained less weight than chicks with natural, bright flanges. However, even if bright flanges start as a signal for detectability, chicks might soon be trying to outcompete each other to have the most detectable mouths, and the trait could become a signal of quality or need.

COLONIAL BIRDS

When birds nest in tightly packed colonies, parent birds face numerous potential complications for directing their parental care to the right chicks. In a crowded colony, simply knowing the exact location of your nest can become slightly more difficult than for solitary-nesting birds. If there are no other nests within 50m or more, as might be the case for many territorial songbirds, a bird is not likely to make a mistake in recognising its nest. However, if the nearest nest is 1–2m away, as in many seabird colonies, the possibility of a mistake rises. Particularly in those cases where the young are mobile soon after hatching, knowing the mere location of the nest may not be enough to allow parents to find their young. Obviously, parents want to know which young are theirs, and so parent–offspring recognition becomes very important. In most well-described cases, vocal signals appear to be the primary trait used in parent–offspring recognition.

Some of the first studies of parent–offspring communication were done in Common Murres (*Uria aalge*, also known as the Common Guillemot) and Laughing Gulls. Field observations noted that parents returning to the colony called as they approached, and young birds oriented towards the calls, but specifically the young appeared to orient towards the calls of their parents. Playback experiments then demonstrated that the young birds could discriminate between the calls of their parents and other adults. Studies on Common Murres found that chicks raised from eggs incubated in a laboratory responded more strongly to adult calls that they had heard while still in the egg, suggesting that the chicks learn the calls of their parents before they hatch. Further playback experiments in a related species, the Thick-billed Murre, found that chicks as young as 3 days old discriminate between the calls of the parents and strangers, or parents and neighbouring adults, but did not discriminate between the calls of neighbouring adults and strangers. These results demonstrate that the

chicks are not merely responding more to calls that are familiar than unfamiliar, but, even amongst familiar calls, they know which calls belong to their parents.

In cases like these, parents find their chicks because the chicks recognise the call of the parents, but often the parents can recognise the calls of chicks as well. Thick-billed Murres recognise the calls of their chicks, as do other members of the Alcidae. For Ancient Murrelets, reciprocal recognition might be particularly useful not because of the density of their colonies, but instead due to their nesting habits. This species nests in burrows or crevices on the ground, often some distance from water. A few days after the eggs hatch, the parents approach the nest at night and call to the chicks from the entrance of the burrow. Under the cover of darkness, the chicks leave the nest and climb through dense vegetation to the sea. Playback experiments have shown that both parents and chicks recognise each other by voice, enabling the chicks to reunite with their parents on the water in the dark.

Razorbills have an interesting variation on the theme of parent-offspring recognition. Both parents take care of the chick while it is at the nest, but only the male provides parental care after the chick leaves the nest and heads out to sea. Playback experiments with Razorbills have shown that chicks recognise the calls of their male parent and males recognise and respond more strongly to the calls of their own chicks. Females, however, show very little response to the calls of chicks, and don't differentiate between the calls of their own chicks and strangers. It might seem odd for the mother not to recognise her chicks, but the calls of her chicks are not a relevant signal to a female Razorbill. As in all forms of communication, when a receiver doesn't benefit from a signal, they tend not to respond.

Studies of swallows provide another clear demonstration of when we should expect parent-offspring recognition to occur, and when it should be absent. Parent recognition of chick calls occurs in colonial Sand Martins and Cliff Swallows, but parental recognition is not seen in closely related, but more solitary nesting, Northern Rough-winged Swallows and Barn Swallows. Only in the colonial species is there a significant risk for parents of confusing their own chicks for others, and only

Sand Martins often nest in large colonies, in burrows they excavate into the soft soil of river banks.

Razorbill males and females have different parenting roles, which results in different patterns of chick recognition.

Parent recognition of chick calls occurs in colonial Cliff Swallows, where dozens or hundreds of birds may construct their mud nests side by side. The more solitary-nesting Barn Swallows do not show parental recognition of chick calls. In the more solitary species, there is little risk of feeding the wrong chicks, and little need for chick recognition abilities to evolve.

in the colonial species do we see the evolution of signals for recognition. The difference in recognition is not just due to parents in the colonial species paying closer attention to their chicks; the calls of Sand Martins and Cliff Swallow chicks are more individually unique than the calls of Northern Rough-winged and Barn Swallows, making it easier to recognise individuals.

Even in the colonial species, the risk of confusing other chicks for your own changes across the nesting cycle. Young Sand Martins are confined to the nest cavity for approximately the first 2 weeks of their lives. During this period, there is little chance that a parent would find a foreign chick in their nest. However, as the chicks become more mobile and start to practise flying, they may end up in the wrong nest. The individually unique chick calls develop at the same time the chicks become mobile. After the chicks leave the nest, Sand Martins and Cliff Swallows often leave their chicks in large groups, known as creches, as the parents go off to forage. The individually unique chick calls will also help the parents relocate their chicks in the crowd.

The use of distinctive vocalisations for parent–offspring recognition is well known, but other signals may play a role as well. In Cliff Swallows, the chicks in the nest also have distinctive facial markings. Human observers can distinguish individual chicks based on these markings, but it's not known if Cliff Swallow parents use this trait to recognise their chicks. It seems possible that they could, but it's harder for researchers to experimentally confirm this. Presenting recordings of calls to a Cliff Swallow is much easier than showing them videos of chick faces.

Lesser Kestrels also illustrate the connection between coloniality and parent–offspring recognition. A study of a Lesser Kestrel colony found that 76% of the nests contained chicks who were unrelated to the parents, adopted from neighbouring nests. Overall, 51% of the nestlings switched nests and were adopted by neighbouring parents. Why would parents so frequently provide parental care to unrelated chicks? One hypothesis is that Lesser Kestrels have not evolved chick recognition abilities because they did not always nest in colonies as they do today. Lesser Kestrels may have originally nested in burrows in cliffs and rock faces. Today, they nest in holes in human-constructed walls and under the tiles on terracotta roofs. In their original habitat, a parent kestrel may have rarely encountered a foreign chick, but on a roof chicks can easily walk from one nest to another,

⬆

Lesser Kestrels usually nest in rocky crevices, either naturally occurring or in human-built structures. They may breed in single pairs or colonially, depending on the availability of nest sites.

➡

Most penguins are colonial breeders. The two largest species, the King (top) and Emperor Penguins (bottom), nest in enormous colonies and build no nest. Instead, the egg is incubated on the feet of the parents. Several weeks after hatching the chicks will assemble into groups known as creches. When parents return to the colony to feed their young, they call out to their chicks, and chicks recognise the calls, enabling chick and parent to reunite under challenging conditions.

particularly in pursuit of an adult with food. Given enough time, we might expect parent–offspring signals to evolve, but, then again, there's no evidence that adopted kestrel chicks or the adoptive parents suffer from the arrangement.

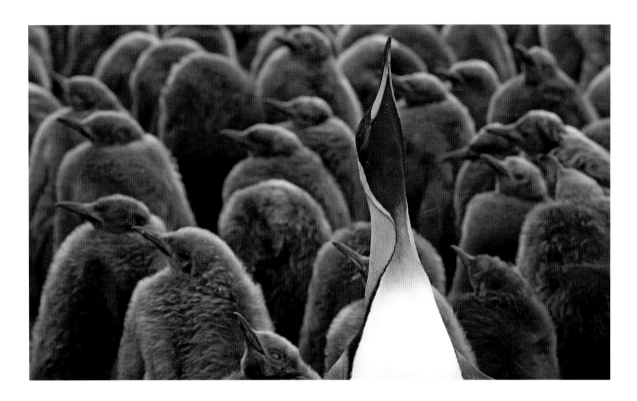

WARNING SIGNALS

For many birds, the world is a dangerous place, with predators lurking around every corner. In response to constant threats, birds have evolved a wide variety of anti-predator behaviours, including both acoustic and visual signals, which will be explored in this chapter.

ALARM CALLS

Calls with an anti-predator function are often referred to by the general term alarm calls, though anti-predator calls can also be divided into smaller categories, such as seet calls, mobbing calls and distress calls.

These different types of calls are categorised by the context in which they are given, and the type or level of predatory threat that exists, but also to some degree by the way they sound.

SEET CALLS

These are high-frequency, narrow-bandwidth calls, typically given by small birds when they detect an avian predator in flight that may be actively hunting for prey. Seet calls cause other birds to freeze or take cover – presumably to make it more difficult for the predator to spot them. The high pitch and narrow bandwidth of seet calls are thought to make them difficult for predators to hear and localise, both because the call will degrade rapidly in the environment and because raptors, such as hawks, kites and other birds of prey, have relatively poor hearing at high frequencies. For these reasons, many small birds from around the world have evolved very similar seet calls.

It is worth noting that many birds produce calls in response to flying raptors that do not have the acoustic structure as described for seet calls above, but still evoke a response to flee or hide. It might be a bit clearer to simply refer to all these types of calls as aerial alarm calls, to point out that they function to alert others to a flying predator.

MOBBING CALLS

Mobbing calls are typically given when a bird has spotted a terrestrial or perched predator. These calls can be short, simple 'chip' notes or longer broad-frequency sounds, that could be described as harsh, rough or raspy. Mobbing calls attract members of the same species (conspecifics) and members of a different species (heterospecifics) in a behaviour known as mobbing. Mobbing may seem counterintuitive – rather than encouraging freezing or hiding, these calls induce birds to run or fly towards danger, with the intention of harassing the predator until it leaves the area.

Sometimes mobbing involves an obvious chase, such as a Northern Mockingbird or a crow chasing a raptor. An ambitious mobber might even physically strike the predator. Other times, such as when chickadees and tits mob an owl, the action involves a close approach and the production of alarm calls, but no physical contact. Other birds are attracted to the mob, but some species will quietly stay at the

periphery. Mobbing sometimes seems to include cases of mistaken identity, as birds will mob non-predators. Mobbing can also go terribly wrong, as the mobbing bird can be killed by the predator. Birders often take advantage of bird mobbing behaviour, either by producing impersonations of chickadee or titmouse mobbing calls, often known as 'pishing', to lure birds in for a closer look, or by listening for mobbing calls which may help the birder find a hidden owl.

Northern Mockingbirds often aggressively mob potential predators, even when the predator might not actually represent a threat, such as this fish-eating Osprey.

This Magpie Shrike is mobbing a highly venomous Boomslang snake, which could be a predatory threat both to the adult bird and its nest.

➡ 'Chip' calls coming from underbrush in Costa Rica might be a pair of White-eared Ground Sparrows contact-calling to stay together, or a pair sounding an alarm.

COMMUNICATING INFORMATION

Although alarm calls may mean simply that danger is present, a growing number of studies indicate that birds can communicate much more complex types of information in their mobbing calls. Mobbing calls may provide valuable information on the presence of danger and the urgency of the situation. However, a simple call might not tell you all that you need to know. For example, there are good reasons why a small bird might want to know if the danger is a small hawk that might pursue a small bird, or a large but less dangerous hawk – to the small bird – that instead targets larger prey. A bird might also want to know if the danger is a hawk attacking from above or a snake attacking from below.

Encoding complex information in alarm calls can be accomplished in a variety of ways. One possibility is to use the same call but vary the rate at which the call is produced. Another possibility is to vary the rate of specific elements within a call, such as the 'D notes' in chickadee calls (see box on pages 130–1). Yet another possibility is to use acoustically distinct calls to indicate different threats. To establish that calls can have such specific meanings, researchers must demonstrate that a bird produces a distinct pattern of call for different stimuli, and demonstrate that other birds, upon hearing the call, will respond to each distinct call in an appropriate way.

The White-eared Ground Sparrow provides an interesting example of the type of communication that is possible with a simple 'chip' mobbing call. The chip notes are given in two distinct contexts: as a contact call as members of a pair forage together, or as a mobbing call when a predator is spotted near their nest. The acoustic structure of each individual chip is the same regardless of the context. The primary difference between their contact call and a mobbing call is the rate at which chips are produced – the faster the chip rate, the more urgent the situation. Signals of this sort are often referred to as 'graded signals'. Chips produced at a faster rate attract other sparrows more quickly, and may even encourage members of other species, such as hummingbirds, wrens, tanagers and warblers to join in and mob a potential predator. Chip notes are quite common among birds around the world, and it seems possible that they serve a similar dual function in many species.

CHICKADEE MOBBING CALLS

Two North American chickadee species, the Carolina Chickadee and the Black-capped Chickadee, have been the subject of many studies on how birds encode information in their alarm calls about predatory threats. Chickadees are named after the sound of their alarm calls, and the 'chick-a-dee' call is very familiar to even casual birders. The 'chick-a-dee' call is made up of several distinct elements, or syllables: chevron-shaped elements (known as A, B and C notes) which make up the introductory 'chick', followed by the broadband D notes. When chickadees are confronted with a predator, they begin to call. The number of D notes produced correlated with the overall size of the predator – smaller predators represent a greater threat and elicit calls with more D notes. A large owl or hawk is less of a threat – either they're not nimble enough to catch a chickadee or perhaps the chickadee is too small a meal for them to bother with – and so elicit calls with fewer D notes.

The Carolina Chickadee is a common visitor to garden birdfeeders throughout the eastern United States.

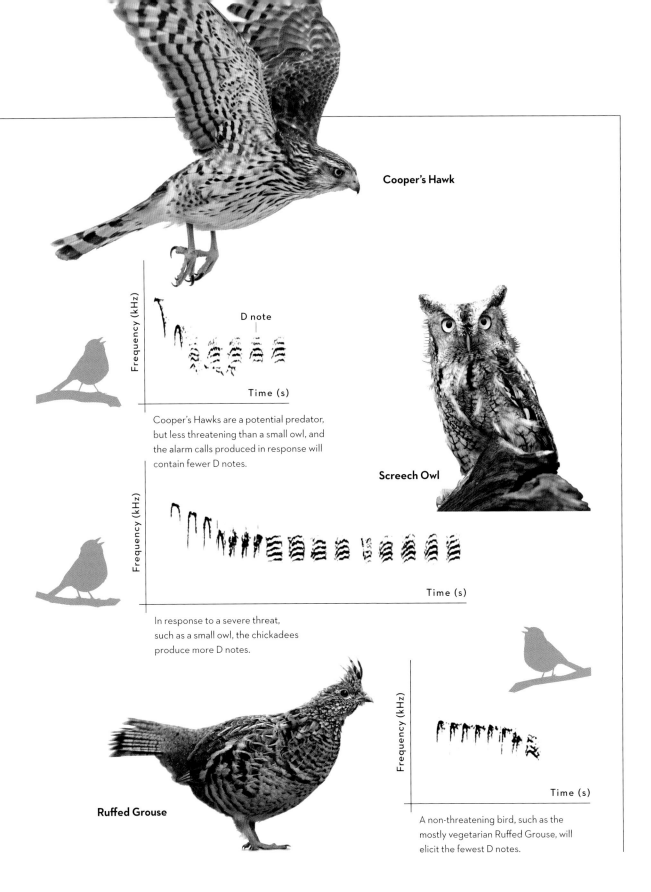

Cooper's Hawk

Frequency (kHz)

D note

Time (s)

Cooper's Hawks are a potential predator, but less threatening than a small owl, and the alarm calls produced in response will contain fewer D notes.

Screech Owl

Frequency (kHz)

Time (s)

In response to a severe threat, such as a small owl, the chickadees produce more D notes.

Frequency (kHz)

Time (s)

A non-threatening bird, such as the mostly vegetarian Ruffed Grouse, will elicit the fewest D notes.

Ruffed Grouse

CALL MEANINGS

Distinct calls with distinct meanings – sometimes called functionally referential signals – were once thought to be the exclusive domain of the human language, but we now know that a wide variety of animals use such signals. For instance, domestic chickens produce different, acoustically distinct alarm calls for aerial predators and terrestrial predators; and Japanese Great Tits produce different alarm calls when they encounter crows and snakes. Experiments have demonstrated that other birds understand the meaning of these alarm calls and respond appropriately. For example, chickens will run for cover upon hearing a recording of an aerial alarm, and stand upright and vigilant upon hearing a terrestrial alarm.

The variation we see in alarm calling, particularly in chickadees, tits and their relatives is so complex, it suggests that these calls hold the potential to encode even more information. Studies have shown that chickadees vary their alarm calling behaviour depending on whether a predator is facing them or not. Chickadees use more A notes in their calls when they spot a flying

Japanese Great Tits have a complex vocal repertoire, and can communicate detailed information about the presence and behaviour of predators.

Chicken alarm calls can indicate different categories of predators, though many of the alarm calls they produce appear to be false alarms, and no threat is actually present.

predator, and more C notes when they are flying themselves. Studies of Japanese Great Tits indicated that the A, B and C notes elicit scanning for danger, and the D notes elicit an approach. It is conceivable that some combination of alarm calls could be indicating the presence of a predator, the type of predator and its behaviour, as well as coordinating the movements of the flock mobbing the intruder.

DISTRESS CALLS

Distress calls are given in the most extreme situation – when an animal has been attacked or captured by a predator. These calls are acoustically distinct from the mobbing or seet calls the same species would produce when a predator is spotted. Distress calls tend to be loud and harsh, broad-frequency sounds that could be described as a 'scream', and, in fact, some authors refer to these calls as 'fear screams'. Such a call is often instantly recognisable as a sign that something truly unpleasant is happening to the screaming bird.

Distress calls have attracted a lot of research attention, even though they are not frequently observed in the wild – most scientists might only experience distress calls from a bird in the hand, caught as part of a research study. Though distress calls might clearly indicate an animal in extreme danger, the intended receiver of distress calls is not always clear. There are a variety of overlapping hypotheses to explain distress calls. One possibility is that the calls function to warn birds of the same species, particularly close relatives, of the presence of a dangerous predator, perhaps enabling the relatives to escape predation. There is

not a lot of evidence that supports this idea. If distress calls are specifically directed at kin, one might expect distress calls to be given more often when close kin are present, but as far as we know, that is not the case. Interestingly, distress calls do not cause others to dive for cover, but instead distress calls often, but not in all cases, attract parents, other family members and flock mates, including members of different species.

The main function of distress calls could be similar to the function of mobbing calls: to attract birds of the same species – conspecifics – so the mob will chase off the predator and allow the calling individual the opportunity to escape. As mentioned above, many birds have a distinct class of alarm calls known as mobbing calls that clearly act to attract conspecifics, but the obvious difference is that distress calls indicate a more severe situation than mobbing calls, and may galvanise a more rapid response.

Another very different set of hypotheses is based on the possibility that the calls are intended for the predator. One idea is that distress calls are a sign that an animal is injured, and attract other opportunistic predators. When the second predator arrives, the ensuing chaos allows the prey to escape. For example, picture a chickadee getting caught by a small Screech Owl. If the distress calls attract a larger predator, such as a Barred Owl, suddenly the Screech Owl is in danger and may give up a potential meal to escape with its own life. This hypothesis has been supported by a variety of observations and experiments. For example, Acorn Woodpeckers, a highly social species that lives in kin groups, regularly produce distress calls when caught by researchers. Those distress calls never attract other woodpeckers, but, in at least one case, the distress calls stimulated a bobcat to run up a tree in search of a meal.

A Eurasian Pygmy Owl is a severe threat to small songbirds like this unfortunate Great Tit.

More evidence for the predator attraction function of distress calls comes from surveys of raptors, such as studies that have used recorded distress calls of Tufted Titmice to attract Red-shouldered Hawks, thus allowing researchers to more accurately count the abundance of the predators. Despite the evidence that distress calls can attract predators, that does not mean that distress calls have evolved to attract predators. For

⬆ ⬈

Tufted Titmice (above) are highly vocal, and their high-pitched distress calls catch the attention of birders and Red-shouldered Hawks (above right) alike.

example, an animal with an injured leg may limp, and a clever predator may focus their efforts on the injured prey. However, no one would argue that limping has evolved to tell predators which prey are injured. We should expect predators to respond to cues of injured prey regardless of whether the cue is intended for them or not.

Finally, the distress call may simply startle or annoy the predator to such a degree that it lets the prey escape. Several studies have used playbacks of recorded distress calls to observe the effect on predators, and have found that some coyotes, racoons or opossums are startled, which could potentially give the calling bird a greater chance to escape. On the other hand, the same studies found that distress calls induced other individual predators to increase their severity of attack. Distress calls might only work to startle the most naive or inexperienced predators, but when a bird is moments from death, anything it can do to help it escape is worth the effort.

⬅

Acorn Woodpeckers produce distress calls when caught by a predator. However, when biologists catch the woodpeckers, some individual birds regularly tend to produce distress calls and others do not, perhaps representing different 'personality' types among the birds.

NON-VOCAL ALARM SOUNDS

Some birds might produce non-vocal acoustic signals with anti-predator functions. Columbiformes, the pigeons and doves, frequently produce flight-related sounds. The Rock Dove and its domestic descendants, feral pigeons, make a noise familiar to most as their wing tips slap together at the top of their upstroke. Is this sound a signal – a trait that has been shaped by evolution to contain information – or an incidental aspect of their flight?

The answer is not truly known. However, pigeons and doves also produce a variety of distinct tonal sounds with their feathers as they take off, such as the 'whistles' that are characteristic of the Mourning Dove of North America and the Crested Pigeon of Australia. Researchers suspected that the Crested Pigeon wing 'whistle' may act as an anti-predator signal, indicating to other doves in the area that a predator has been detected, and they should take flight immediately.

To understand if the whistle is a signal, first we might want to establish if the sound of a dove taking off in alarm is different from a dove taking off in a more leisurely fashion. Researchers attracted Crested Pigeons to feeding stations and recorded their flight sounds. For the leisurely non-alarm sounds, they simply recorded the whistle of a pigeon spontaneously taking off and leaving the feeding station. For the potential alarm signals, the researchers tossed a model of a hawk in the direction of the feeder and recorded the sound of the pigeon taking off. Analysis of the recordings showed that alarm whistles are both louder and more rapid in tempo than non-alarm whistles. This alone does not establish that

Pigeons and doves make a variety of non-vocal sounds as they take flight, from the simple wing 'claps' of the Rock Dove (left) to wing whistles in the Mourning Dove (right).

The Crested Pigeon of Australia is the best studied example of the alarm function of non-vocal sounds in pigeons.

the sound is a signal, as we might expect that flights in response to a predator would be more energetic, and this alone might result in acoustic differences.

A second set of experiments used sound playback to ask if the different whistles appear to contain information. Recordings of alarm and non-alarm whistles were played to crested pigeon flocks, and, in about 70% of cases, the pigeon flock immediately took off in response to the alarm whistles, but no flocks took flight in response to non-alarm whistles. This result suggests that the sound could contain information about danger, but is there any evidence that natural selection has specifically shaped the pigeons' wing to produce this sound? This evidence comes from the shape of the primary flight feathers. The Crested Pigeons' eighth primary feather is noticeably different, about half the width of the feathers on either side of it – this uniquely shaped eighth primary is not found in other species of Australian doves and pigeons. By experimentally removing the eighth primary, researchers showed that this highly modified feather was responsible for the high note of the alarm whistle. Another series of playback experiments were performed which compared responses to alarm whistles of birds whose eighth primary was removed with alarm whistles from birds with normal wings. It was found that the alarm whistle without the eighth primary feather was less effective in causing birds to take flight. It appears that this key modification of one of the flight feathers took a sound and turned it into a signal.

EAVESDROPPING

While most bird songs are only of interest to other members of the same species, an interesting aspect of bird alarm calls is that this communication system crosses species boundaries. Perhaps not surprisingly, if a chickadee spots a dangerous predator, that information is also valuable to scores of small birds of other species who might be vulnerable to the same threat, and many birds pay close attention to the seet, distress and mobbing calls produced by other species.

Non-birds, such as iguanas, chipmunks, lemurs and monkeys will also respond to the alarm calls of birds. Even the finer details of a Black-capped Chickadee call that indicate the presence of a small hawk or a big hawk are understood by other bird species, such as Red-breasted Nuthatches.

Birds may recognise alarm calls through a variety of mechanisms. As mentioned above, seet, mobbing and distress calls tend to have acoustic characteristics that are very similar from one species to the next, and so the alarm calls of species A and species B may sound so similar that mutual recognition of the call is straightforward. There are a variety of possible explanations for why the alarm calls of different species sound so similar. Perhaps the seet calls of so many birds are similar because, in each case, natural selection has favoured the evolution of calls that are difficult for predators to localise. Perhaps the distress calls of many animals are similar because the harsh, screaming sounds work well to startle a wide variety of predators. Mobbing calls may be similar because mobbing is beneficial to all the species involved, and so natural selection has favoured a convergence of these calls to elicit larger, more effective mobs.

Red-breasted Nuthatches mob more strongly when they hear a chickadee alarm call indicating a dangerous small owl than the call type that indicates a large owl.

⊙ ⊙ The Superb Fairywren
(left) and the White-browed
Scrubwren (below) have a
shared interest in detecting
predators, and where their
geographic ranges overlap,
they appear to learn to
understand each other's
alarm calls.

Birds may also learn to recognise alarm calls that are quite different from their own. Learning alarm calls could be particularly beneficial when the calls of different species are not acoustically similar, and yet there is valuable information to be gained through eavesdropping. A fascinating case of learned recognition appears to occur in Superb Fairywrens, which produce aerial alarm calls when they spot flying raptors. Another small species of bird, the White-browed Scrubwren, is found in many of the same Australian forests and produces very similar-sounding alarm calls. However, experiments have shown that the fairywrens respond to scrubwren alarm calls only in areas where the ranges of the two species overlap. Superb Fairywrens at the western edge of their range, in an area where scrubwrens do not occur, do not respond to the scrubwren alarm calls. At a finer geographic scale, Superb Fairywrens will respond to the aerial alarm calls of Noisy Miners, but only in locations where the territories of fairywrens and miners overlap. A fairywren less than a kilometre away, in a location that happens not to have a miner territory, will not respond to miner alarm calls. In these cases, it appears that the fairywrens only respond to alarm calls of the miners when they have experienced them, and have learned to respond to the Noisy Miner signals.

Noisy Miners may be noisy neighbours, but their alarm calls are useful information to fairywrens in overlapping territories.

ALARM CALL RELIABILITY

Despite the importance of alarm calls to both the signaller and the receiver, and the fact that the calls can carry very specific information about predatory threats, alarm calls are notoriously unreliable. Birdwatchers can use bird alarm calls to detect predators such as owls, but a birder who follows every alarm call they hear will often be left wondering what the birds were fussing about.

In the studies that established that domestic chickens have distinct aerial and terrestrial alarm calls, 45% of the times when the chickens gave an aerial alarm, the researchers could not detect any object that should have caused the birds to alarm. Even when the researchers noted that an object in the sky correlated with the chickens' aerial alarm, almost 40% of the time the object was something which was unlikely to represent a threat, ranging from flying ducks to small songbirds, and even insects or a leaf. In other words, most alarm calls appear to be 'false alarms', when an alarm call is produced but no predator appears to be present.

This pattern raises two questions: why are so many false alarms produced, and why do birds continue to respond to false alarms? The second question might be easier to answer. Receivers might respond to false alarms due to 'adaptive gullibility'. Alarm calls provide important information about potentially lethal threats and the benefits to responding to alarm calls are extremely high. On the other hand, the costs of responding to false alarms are probably low, so the best thing for a receiver to do is to respond to the alarm whether it is true or false.

But why are so many false alarms produced? One possibility is that the false alarms are simply mistakes. Just as it pays to react to an alarm call even if it might be false, it might pay to give an alarm call in response to a fast-moving object in the sky, even if it later proves to be a dove rather than a hawk. Another possibility is that some individuals might be particularly jumpy or excitable, and thus more prone to giving alarm calls. In many animals, young individuals are known to produce more false alarms than older individuals. Even within an age class, tremendous variation in 'personality' has been described in many bird species, and a nervous bird might produce a lot of false alarms while a calm bird might give more reliable alarm calls.

DECEPTION

There are also situations when producing false alarms is advantageous, such as when strategic use of false alarms of aerial calls can drive competitors away from feeding opportunities. Great Tits form dominance hierarchies during the winter which influence their ability to use good feeding sites. Subordinate individuals will use false alarms to scare off both dominant and subordinate birds of the same species. Dominant birds also will use false alarms to scare off other dominant birds from a feeder. Deception is not necessary for a dominant bird to scare off a subordinate.

Deceptive use of alarm calls can also occur between species. One well known example is the White-winged Shrike-tanager, which lives in mixed-species flocks that forage together in the Amazon rainforest. One benefit of living in a big flock may be that there are many eyes on the lookout for predators, and the shrike-tanagers regularly give alarm calls when they detect a hawk. However, one important cost of group living is that the members of the mixed flock are often competitors for the same food. If another species in the flock, such as an antbird, flushes a large insect, the shrike-tanager gives an alarm, and when the antbird freezes or dives for cover, the shrike-tanager then has a chance to grab the prey.

In these cases of deception, false alarms result in missed foraging opportunities, and the costs of responding to the false alarm are increased. Under these circumstances, we might expect that receivers would become more sceptical, and less gullible, about alarm calls. We might then expect the false alarm producers to become cleverer in their use of false alarms, which sets up the possibility of an arms race of sorts between the false alarm producers and receivers. Fork-tailed Drongos may represent a further step in that arms race. The drongos use a combination of false alarms and mimicry to steal food from flock mates. The drongos produce drongo-specific alarm calls, but they also mimic the alarm calls of other species such as the Pied Babbler. If a babbler finds food, the drongo can produce its own call or mimic the babbler's alarm call.

By responding to a White-winged Shrike-tanager alarm call, a Black-faced Antbird runs the risk of being duped out of a meal, but the risk of predation is even greater.

Both calls will startle the babbler momentarily, but the mimicked babbler alarm causes them to stop feeding for a longer period. The babblers may still habituate, or reduce their response, to alarm calls after repeated harassment, and the drongos will respond by switching the type of alarm call they are using. Natural selection may continue to produce adaptations and counter-adaptations in drongos and babblers as long as the threat of predation continues to hang over their heads.

Several species of drongos in Asia and Africa, like the Fork-tailed Drongo pictured here, have been documented using false alarm calls to steal food from other species.

The Pied Babbler is a regular victim of the Fork-tailed Drongo's trickery.

This Killdeer is performing a broken-wing display to distract a predator, but a clever predator might realise this action means a nest of young birds are nearby.

VISUAL DISPLAYS

As well as acoustic signals aimed at the predator, many birds also perform conspicuous visual displays when they detect a predator, and the most well-known anti-predator displays clearly have the predator as the intended audience, and seem to be intended to attract, distract or startle the attacker.

DISTRACTION

Perhaps the most well-known anti-predator displays are broken-wing displays. In these, the bird feigns injury by running along the ground in an awkward and exaggerated fashion, often with their wings extended and listing to one side. Having caught the predator's attention, the parent bird then attempts to lead it away from vulnerable eggs or young. These displays are performed by many species, particularly shorebirds, but also grouse, nightjars and songbirds.

Though these displays are widespread, they are not always particularly effective. This is another situation in which selection is acting strongly on the signaller (the parent bird) to deceive the predator, but selection should also be acting strongly on the receiver (the predator) to avoid being deceived. Observations of

This female Common Nighthawk is performing a distraction display near her nest.

grouse using distraction displays against foxes have noted that the foxes sometimes ignore the parent feigning injury, and instead the fox searches the immediate area for chicks. The fox has somehow figured out that the distraction display provides no information about the presence of an injured adult, and instead is more accurate as a predictor of vulnerable chicks being nearby. Much like distress calls, distraction displays may only work against the most naive predators.

DECEPTION

In response to predatory threats, many birds undergo a striking transformation in which they raise their feathers, spread their wings and hiss or snap their bills. This sort of communication is usually thought of as a bluff – the bird makes itself appear to be much larger than it really is, and the display might convince a predator that the bird is too large and intimidating to be prey. Interestingly, threat displays of this sort are common among large birds such as owls, which have some genuine weaponry which could be used defensively to back up a bluff. These displays may also be an attempt to startle, particularly in bird species such as the Sunbittern, which spreads its wings to reveal colourful feathers, which some have interpreted as eyespots.

A Sunbittern spreads its wings in a display that is directed towards both potential predators and competitors.

In some smaller birds, threat displays seem to work using mimicry. The Eurasian and African woodpeckers known as wrynecks are named for their head-twisting threat display. When disturbed at their nest, or caught by a bird researcher, the birds sway their heads and hiss, giving the impression of a snake. A predator peering into a dark tree cavity in which these birds nest might easily be convinced that they have encountered a snake and retreat. Other cavity-nesting birds, such as chickadees and titmice, produce similar snake-mimicking displays, with sinuous movements, hisses and even lunging or striking with their bills and thumping the sides of the nest cavity with their wings. Even for a researcher who knows that the sounds are coming from a chickadee and not a snake, the overall display can be quite surprising and intimidating.

PURSUIT-DETERRENT DISPLAYS

The threat displays described above work because they involve surprise and deception. Another category of anti-predator display, often known as predator-deterrent or pursuit-deterrent display, may work because it contains reliable information. Pursuit-deterrent displays are a bit more mysterious than distraction or threat displays, because it is often not immediately clear that the signal is meant for a predator. It takes careful, repeated observations to confirm that the display is given primarily in the presence of predators. Even after the association with predators is confirmed, one must confirm that the display is not intended to inform conspecifics such as a mate or offspring of the presence of the predator.

Motmots are well known for their long racket-tipped tails. The tails look like a display ornament of some sort and can be wagged from side to side in an exaggerated manner. The tails are found on both sexes, however, which suggests that they don't function in mate attraction, at least not in the typical pattern of males attracting females. Many observers over the years have noted that motmots typically keep their tails still, but will begin to wag them when approached by a human. Careful observations and experiments have revealed that motmots wag their tails in the presence of predators, and they will do so regardless of whether other motmots, such as mates, are present. If tail-wagging is performed in the presence of a predator, with no other conspecifics around, it suggests that this signal is aimed at the predator. Pursuit-deterrent signals are thought to function primarily by informing the predator that they have been spotted, that the potential prey is on alert, and so the element of surprise is gone. Additionally, the signal could be an honest signal of quality. Just like sexual displays could inform a female which male is in the best condition, a pursuit-deterrent signal could inform the predator that the potential prey is in good condition and has a greater likelihood of escape.

A handful of other bird species are thought to use pursuit-deterrent displays. A common theme among most of them is that the display involves tail-wagging or flashing colours in tails, wings or crests.

The odd display of the Wryneck may be mimicry of a snake, a phenomenon seen in several other cavity-nesting birds including chickadees and their relatives.

Several species of birds, including Turquoise-browed Motmot (right) and trogons, perform tail-wagging displays that seem to be directed towards predators, potentially as a pursuit-deterrent display.

GROUP LIFE

All communication is a social act, even if the communication only mediates the interactions of two individuals, such as mates, or fighting neighbours. Birds often live in large groups, ranging in size and type from a dozen birds in an extended family unit to many hundreds of birds in a flock. In this chapter, we look at signals that have evolved to enable birds to show their membership of a family or a flock, and to help coordinate the activity of their groups.

FLOCKS

Birds form flocks for a variety of reasons. For instance, they may gain safety in numbers by increasing the number of vigilant eyes in the group on the lookout for predators. Flocks could also produce a dilution effect, so that even though a predator might take one bird, the overall chances of any one individual being taken and killed are decreased. Another benefit to flock life might be more efficient foraging, with individuals following a group of birds to find food. At the same time, there are potential costs to group life, such as increased visibility to predators, and the presence of more competitors for food.

The well-documented benefits of group life often suggest that flocks consist primarily of unattached individuals which are drawn together for entirely selfish reasons. However, flocks may also consist of a mixture of mated pairs and family groups, which all have a shared interest in gaining benefits from the flock. Jackdaws are monogamous and have long-term pair bonds, but they are also highly social and join large flocks in the winter. Recent studies of Jackdaws have noted that, within a flying flock, there are duos – potentially mated pairs – that tend to fly closer together.

Tundra Swans leave their breeding grounds in the high Arctic and migrate to refuges on the coast of North Carolina. Within the flocks, which can number in the thousands, are many individual families, consisting of mated pairs and their offspring. Emperor Penguins breed in huge colonies on Antarctic ice, with relatively few landmarks to help them find their nests. When mates swap roles during incubation and chick-rearing, they find each other within a crowd of tens of thousands of other penguins. Keeping track of a partner and offspring within a swirling mass of individuals presents a problem. How do birds keep from getting separated? If they become separated, how do they find each other again, and how do they maintain a healthy relationship? The answer is communication.

 This flock of jackdaws may look chaotic, but the movement of the flock is more coordinated than it might first seem.

 Tundra Swans migrate in flocks of thousands, and within these flocks, parents and their offspring stick together. Tundra Swan flocks make for an amazing sight, and also an amazing sound, as thousands of birds call to each other, both on the ground and in the air.

In a crowded colony on featureless ice, communication is key for Emperor Penguins to find their mates and their chicks.

↑

Even a photograph can vividly
conjure up the sound of a flock of
Canada Geese flying overhead.
The sounds we hear are the
contact calls, which function to
keep families together within
the flock, as well as help the
entire flock to stay together.

CONTACT CALLS

**Bird flocks are in constant communication – they are noisy, they honk,
trumpet and chatter. When a skein of geese or a flock of goldfinches,
siskins or crossbills passes overhead, one is often first alerted to their
presence by their voices. What is the function of all these sounds?**

Many calls given by birds in flocks, particularly those calls that are not
clearly alarm calls, are thrown together into a category known as contact
calls. These are not always acoustically distinct from other types of calls:
chickadees produce their namesake calls as contact calls as they move
through the forest in a winter flock, and also as alarm calls when they spot
a predator. Contact calls can be given in the smallest social groups, such
as by mates foraging together on their territory, or in large, mobile flocks.

Compared to the amount of research on alarm calls, contact calls
have been relatively poorly covered, and, in many cases, the functions of
contact calls have not truly been experimentally verified. However, it is
likely that contact calls allow birds to do two related things: coordinate
movements of a group, and recognise preferred social partners.

COORDINATING GROUP MOVEMENTS

Perhaps the simplest function of contact calls is to encourage group cohesion – in other words, using these signals helps to keep the flock as a flock. Contact calls might do little more than let other individuals in your group know where you are and if you are moving. If receivers respond to the signals by staying close, and moving when others move, then the flock sticks together. In a small group, birds can communicate directly about their location and their movements, and so flock cohesion may be relatively easy to achieve.

Green Woodhoopoes live in extended family groups which contain a dominant breeding pair and several non-breeding subordinates. A group tends to forage together as a flock while moving through their territory. When individuals head off towards a new location to forage, they often produce a call, and individuals who call are much more likely to be followed by other members of the group. If a dominant individual leaves and is not followed, whether or not they called before they left, they often go back to the group and call again – the dominant birds expect to be followed. When subordinates leave, if they called first, they are likely to return to the flock and call again. Subordinates who leave silently and are not followed tend to continue foraging alone.

Green Wood Hoopoes live in large, loud and chattering family groups.

Group-living Pale-winged Trumpeters produce a wide variety of vocalisations, including territorial songs and alarm calls. But they also produce distinct 'mew' calls as the group forages in the dense forest understorey. Mew calls are produced by individuals when they get visually separated from the rest of the group, and playback recordings reveal that hearing a mew call causes other group members to stop what they are doing, and they may produce a 'grunt' call in return. If the mew signals: 'Where are you?', the grunt signals: 'Over here!'

In larger flocks, it becomes harder or even impossible for all birds to communicate and coordinate the movement of the group. It is likely that communication is limited to closely spaced individuals, and yet the entire flock may appear to move as a coordinated whole. In migrating groups of swans, geese and cranes, various pre-flight behaviours have been described that might function to achieve consensus as an entire flock is getting ready to undertake a long flight. Many of these signals are postural, described as head-bobbing and head-shaking in Bewick's and Whooper Swans, or head-tossing in Canada Geese. Sandhill Cranes engage in a ritualised neck-stretch display before flight, which may include spreading the wings and running. The cranes' behaviours are not merely necessary warm-up activities, as the birds can

The 'mew' calls of Pale-winged Trumpeters help the group stick together as they move through dense forest, and these calls appear to be individually distinctive, meaning all of them will know when an individual is missing.

Large waterfowl like these Whooper Swans (left) and Canada Geese (right) perform head bobbing displays before they take flight. The displays allow the signaller to indicate that they intend to take flight, and the receiver can reciprocate, ensuring that pairs take off together.

leap into flight without performing the behaviour, such as when confronted by a predator. Thus, these behaviours may truly act as signals of impending flight, with the hope that nearby receivers, particularly the mate or offspring of the signaller, will be encouraged to take flight as well. For large flocks of birds, we still don't know how the movements of an entire flock are coordinated and how much communication is involved. It may be that many subgroups are responding to the same stimulus to move, or it may be that pre-flight signals are 'contagious', and eventually a critical mass of birds are stimulated to take flight.

Sandhill Cranes often travel
as pairs or family groups,
and both calls and postural
displays help coordinate
movement.

MIGRATORY FLIGHT CALLS

Many species of migratory songbirds produce distinct
calls at night during their flights. The function of these
calls is not known. Most songbirds do not migrate in
family groups, so it is unlikely that these calls are intended
for close relatives. But as with so many other calls, they
are presumed to aid in keeping the flock together.

A disturbing piece of evidence for the function
of these calls comes from a study of birds that collided
with buildings during migration. A huge number of birds
die in collisions with high-rise buildings, and, in many
locations, scientists collect and catalogue the unfortunate
victims. The pattern of deaths have indicated that the
species that produced more nocturnal flight calls
tended to collide with buildings more frequently than
the species not known for using nocturnal calls. This
suggests that the flight calls do help keep flocks
together, but when artificial lights and tall buildings are
involved, flock cohesion can have terrible consequences.

MATE RECOGNITION

Contact calls also serve an important role in allowing recognition of specific social partners, and this function has been demonstrated in many flocking species. Perhaps the most important social partner to keep track of is a mate. American Goldfinches frequently forage at a long distance from their nest, and, in the winter, they form large flocks. It would seem to be useful for American Goldfinches to keep track of each other's movements, and the constant flight calling of a flock of goldfinches suggests that a large amount of communication is taking place.

While it is difficult to perform experiments that would accurately simulate communication in a moving flock, many studies of birds in captivity have shown that flocking birds such as American Goldfinches, Silvereyes and Zebra Finches recognise the contact calls of their mates. In American Goldfinches, Pine Siskins, crossbills and Eurasian Siskins, the contact calls of mates appear to converge over time and to sound very similar, which may facilitate mate recognition. Experiments on captive Zebra Finches have shown that calling rates increase when a mated pair is visually isolated from each other, providing further evidence for the role of contact calls in keeping track of a mate.

(Top) Silvereye contact calls help pairs stick together as they forage in flocks, and researchers have shown that in noisy urban environments, they produce higher-frequency calls to be heard above the noise.

Finches such as American Goldfinch (above) and Eurasian Siskin (pair below) are often first detected by their constant contact calling both while foraging and in flight.

This young Western Bluebird may someday be feeding its own offspring or younger siblings – in both cases it helps promote survival of its own genes.

FAMILY RECOGNITION

One of the key prerequisites that allow for complex social behaviour is the ability to recognise other individuals. In territorial communication between two birds, the ability to recognise the vocal signals of neighbours allows for reduced aggression between neighbours (the 'dear enemy' effect, see pages 86, 91 and 92), even if the songs of territorial birds don't exist primarily to allow for recognition. As groups grow larger, the need to recognise group members can become even more important, and thus we see the evolution of signals that allow birds to show their membership in a family or in a flock.

Of the world's 10,000-plus species of birds, a few hundred species are known as cooperative breeders, and live in extended family groups. Cooperative breeding is a very interesting social system, in which there is a primary breeding pair and young birds from previous breeding seasons. Staying at home, rather than moving out and starting a territory of one's own, can be valuable if individuals that stay on their natal territory gain a chance to inherit it from their parents, or if they stay at home to act as 'helpers' and assist their parents in raising a new generation of siblings. Raising new siblings can be an effective way to ensure copies of the individual's genes reach the next generation, because, in diploid organisms – including mammals and birds, which have paired chromosomes, one from each parent – the bird shares as much genetic material with its full siblings as with its own offspring. So, from an evolutionary point of view, helping to produce an extra sibling can be just

as valuable as producing one offspring. Helping in this way only works if the bird can direct its helpful behaviour towards kin. Helping unrelated individuals produce more offspring does not help the bird's genes spread, and so it needs to be able to discriminate kin from non-kin, and, in many cases, signals exist which aid in making that identification.

Western Bluebirds typically breed as pairs, but males frequently disperse short distances and settle on nearby territories, resulting in neighbourhoods where the neighbours may be closely related. If a male is unsuccessful breeding on his own territory, he may return to his parents' or a brother's territory to help raise siblings, or nieces and nephews. Clearly, being able to recognise kin would be important, to ensure that the bird is helping in the right place, and playback experiments have shown that Western Bluebirds recognise kin by their songs. However, these songs are also used in mate attraction and territory defence, so it is probably not the case that songs exist to give a signal of kinship. Instead, Western Bluebirds may use the same ability to learn songs to distinguish neighbours from strangers found in many territorial birds, but in this case that ability is applied to distinguishing between kin and non-kin. In a parallel to the dear enemy effect, the bluebirds respond more aggressively to the songs of non-kin. It makes sense that a bird might be less aggressive towards kin, as kin are likely to represent less of a threat. It remains to be seen whether the ability to recognise kin by song helps bluebirds decide who to help and who not to help.

In highly social Stripe-backed
Wren groups, learned calls
allow individuals to recognise
the sex and family lineage of
other birds.

Stripe-backed Wrens live in extended family groups in the savannah forests of Colombia and Venezuela. They produce repertoires of calls known as 'WAY' calls, which are apparently learned from same-sex older relatives. Males in a group share the same call types, and females in a group share their call types, but the male and female repertoires are different. On occasions when males disperse to breed on neighbouring territories, they take their calls with them, which means that related males, even on different territories, will share the same calls. Playback experiments have demonstrated that WAY calls are frequently used during territorial skirmishes between groups, and that males can discriminate between the calls of group members and non-group members. It is possible that these calls act as a signature of group membership and help members of groups identify each other during fights.

Bell Miners are another cooperatively breeding bird using calls that primarily function as kin recognition signals. Bell Miners often live in large colonies, potentially of hundreds of individuals, which can consist of multiple cooperatively breeding extended families all defending a shared territory, so they are in close contact with both kin and non-kin. In this social system, there is a reasonable chance for individuals to help non-kin by mistake, but studies have shown that closeness of kinship is a strong predictor of helping behaviour. Bell Miners produce mew calls that are not used in a territorial context, but instead are given by birds visiting the nest. Mew calls are more similar between kin than non-kin, and, in fact, there is a linear relationship between call similarity and relatedness, which means that call similarity could be used as a very accurate signal of relatedness. Helping behaviour is also strongly correlated with call similarity, and helpers provide food for young at greater rates when their calls are more like the calls of the breeding male at a given nest. These results suggest that Bell Miners are using call similarity to make accurate decisions about who to help and how much help to offer.

The Bell Miner is named for its characteristic bell-like call, but it uses another vocalisation, the 'mew' call, to assess kinship in its social groups.

However, there could be situations which favour deceptive use of kin signals. If Western Bluebirds use song to recognise kin, and related individuals show less territorial aggression to each other, then it would benefit non-kin to fake their kinship status and thus receive lower levels of aggression. In the case of the Bell Miners, where helpers may use the similarity between their calls and a breeding male to make helping decisions, deceptive use of kin recognition calls by breeding males could potentially result in birds offering to assist non-kin, which benefits the breeding male, but is costly to the misdirected helper. On the other hand, deceptive use of calls might also result in actual kin decreasing their help.

Regardless of whether deceptive use of kin signals could be favoured, a bird could still make costly mistakes in recognising kin, and offer help to those that

The Long-tailed Tit is another highly social bird in which learned calls result in call similarity between group members, which allows birds to make sure that helping behaviour is directed towards relatives.

Chestnut-crowned Babblers recognise group members based on individually distinct calls, rather than group-specific calls.

are not genetically related. How is the reliability of kin recognition signals ensured? One possibility is that the signals – either songs or calls – used in kin recognition are learned by young birds from their older relatives. In this way, family vocal signatures could be passed from parent to offspring and from generation to generation. In Stripe-backed Wrens, Long-tailed Tits and others, calls appear to be learned. Learned kin recognition signals work particularly well when learning is restricted to young birds (often true for vocal signals), and when young birds are rarely in the presence of unrelated individuals.

Another possible way that kin could share more similar calls is if there is a genetic mechanism that controls call structure directly, or if genes control the shape of the vocal tract in some way that influences call structure. Of course, kin will share more similar genes, and if those genes influence call structure, then they would likely share similar call structures as well. In Bell Miners, it has been argued that the calls are not learned, or at least, they do not appear to change across the development of individuals.

Finally, kin recognition could be achieved in cases where there is no vocal signature, or no close vocal similarity between kin. In Western Bluebirds, kin recognition by song occurs, but there is no evidence that the songs of kin are more similar than the songs of non-kin. Chestnut-crowned Babblers, another cooperative breeder, use calls that are individually distinctive, and respond differently to calls of kin and non-kin, but there is no evidence of greater similarity of the calls of family members. Instead, associative learning could allow birds to connect specific individuals with specific calls, and thus recognise familiar and unfamiliar individuals, and if the most familiar individuals are closely related, then kin recognition can result.

← While the female is incubating, the male Green-rumped Parrotlet leaves the nest area to forage, then produces contact calls, which the female can recognise, on his return.

↓ Orange-fronted Parakeets respond more strongly to geographically local calls rather than those that are more distant, and can also change their own calls to match the calls of others, which allows them to join unfamiliar flocks.

→ Red-and-green Macaws, like many parrots, travel in large noisy flocks as they head to foraging locations and clay licks.

↘ For Budgerigars, using contact calls to recognise your mate might be particularly valuable when you travel in swirling flocks of hundreds or sometimes thousands of individuals.

GROUP RECOGNITION

Parrots are well known for their vocal ability. Most species of parrots produce loud contact calls that can be heard as flocks fly overhead and forage in the forest canopy, and they can probably recognise the calls of at least some other individuals in their group. Budgerigars, Green-rumped Parrotlets and Spectacled Parrotlets can recognise their mates by their contact calls. Sub-adult Spectacled Parrotlets discriminate between the calls of siblings and strangers. Other species of parrots likely use contact calls to distinguish between members of their flock and strangers, such as Orange-fronted Conures, which can discriminate between familiar and unfamiliar calls.

Studies of the Brown-throated Parakeet may help us understand how parrots use contact calls in nature. Parrots, in general, tend to travel widely to forage in large flocks, rather than remain in a local territory or home range. When Brown-throated Parakeets find a good food source, they settle into the forest canopy, and, as they forage, other groups of conures fly overhead. The groups overhead sometimes produce contact calls. The foraging group in the canopy may produce contact calls in response or they may stay quiet. A group flying overhead is only called to when they are calling as well, and the overflying group is much more likely to join the foraging group when the foragers produce contact calls.

Playback experiments show that calls alone are enough to encourage an overflying group to land in the canopy. The contact calls seem to act as an invitation to certain groups to join the feast. But why would one flock want to share their food with others? In truth, the answer still is not clear. Increasing the size of the group may have anti-predator benefits, but a more frequently cited hypothesis is that parrots interact with other flocks to increase foraging efficiency. Other flocks may have knowledge about the location of more food resources. Perhaps inviting other flocks to feed may have a future payoff, if, at a later date, the roles are reversed in a form of reciprocal altruism. Being able to recognise individuals may allow a bird or a flock to know which other individuals or flocks have shared food finds in the past and which ones have not.

Red Crossbills offer another very interesting case study of the value of communicating flock membership. Crossbills are highly social birds which occur in flocks throughout the year. They have very restricted diets, specialising on the seeds of conifers such as spruces, pines and hemlock, and they extract the seeds from the tree cones using their unique bills which cross at the tip. As crossbill flocks move through a forest, or even across entire landscapes and continents looking for mature cones, they constantly produce contact calls, both while perched in trees and in flight. Different populations of crossbills specialise in different kinds of conifers. The different populations have distinct morphological adaptations to their favoured conifer (such as a larger bill

size for pine specialists, and a smaller bill for hemlock specialists), and have distinct contact calls as well. In North America alone, there are at least ten different call types, but the geographic ranges of the call-type populations can overlap, meaning that individuals of different call types can be found in the same place.

The Brown-throated Parakeet, like most
species of parrot, is not territorial, and forms
'fission-fusion' flocks, which individuals
regularly join and leave as they forage
during a single day.

⬅

Red Crossbills are one of just a few species of birds with bilaterally asymmetrical beaks, which they use to pry open conifer cones to get at the seeds. The bills can cross in either direction (top to the left or top to the right).

⬇

Call convergence in Galahs may help them to quickly negotiate social interactions, and help them to find a foraging flock to join, or maybe just a playmate.

Crossbills' calls may help them keep track of their mates in the crowd, but the calls might also help a crossbill to recognise a flock of the right call type, so that they can find food as well as an appropriate mate. Studies in multiple species of crossbills in both North America and Europe have found assortative mating, or a preference of individuals of a given call type to mate with individuals of the same type. The correlation between foraging adaptations, call types and mating preferences has the potential to allow further morphological and genetic divergence between crossbill types, and could eventually cause the ten North American call types of Red Crossbills to diverge into different species.

In the case of Red Crossbills, the reliability of contact calls as signals of group membership may arise because contact calls are learned early in life, typically by young birds still in the nest. Cross-fostering experiments show that crossbills learn their calls from whoever raises them, rather than representing an innate signal. Crossbill calls remain relatively stable over their lifetime, though the calls of members of a pair may converge to resemble each other, but no crossbill ever produces the call of more than one of the ten call types known in North America. The call that a red crossbill learns early in life should reliably signal which call type the bird belongs to, and, generation after generation, calls learned in this way should continue to correlate with what foraging adaptations the individual possesses. It seems highly unlikely that any individual would ever be favoured to produce an unreliable, or 'wrong' contact call, as that would lead to them attracting mates and social partners with different foraging adaptations. If a pine specialist followed a flock of hemlock specialists, the pine specialist would be led to an inappropriate food source. If a pine specialist mated with a hemlock specialist, the pair would end up producing offspring poorly suited for either food source.

Parrots also learn their contact calls early in life. Green-rumped Parrotlets learn individually distinct contact calls from their parents while still in the nest. At the same time, parrots are well known for being great imitators, and some species of parrots can modify their contact calls later in life. For example, Yellow-naped Amazons, which occur throughout Costa Rica, have geographically variable contact calls, and there are three distinct contact call dialects, each found in roosts in different parts of the country. Individuals from locations near the boundary of two dialects might have two contact calls, one for each dialect. Researchers experimentally transported adults and juveniles from one dialect area to another, and found that juveniles will learn the local contact call, whereas adults will not.

Parrots can also modify their contact calls based on social context. Female Budgerigars prefer to mate with males with similar contact calls, but male Budgerigars will mimic the contact calls of a female during courtship. Two other species of parrots, Orange-fronted Conures and Galahs, can quickly modify their contact calls to match (or not match) the contact calls of other individuals they hear. The mimicry skills of parrots and their ability to modify their contact calls would make sense if the primary function of their contact calls is to socially interact with other individuals, such as mates, family members or other flocks.

FOOD CALLS

Another form of social communication involves the production of specific, acoustically distinct calls, known as food calls in the context of foraging.

A lone House Sparrow will produce a distinct call, different from a general contact call, when they discover food, and this call attracts other House Sparrows, resulting in a group foraging together. Chickens produce a distinct call when they discover food, and playback experiments demonstrate that the chicken food call is not merely a contact call that encourages flock formation, but a functionally referential call, similar to what is seen with some alarm calls – the call genuinely signals the presence of food as opposed to merely acting as a signal to assemble. Playback experiments show that when other hungry chickens hear recordings of the food call, they aren't simply attracted to the call, but they start actively searching the ground. A chicken that already has food doesn't respond.

Calls that signal the presence of food and function to attract other birds of the same species may seem counterintuitive, since the result could be that less food is available to the signaller. There must be benefits for the caller to produce food calls, but the specific benefits

may vary from case to case. Cliff Swallows, which breed in large colonies, produce a call when they find the flying insect swarms on which they prey, and playback experiments have shown that other Cliff Swallows are attracted to this call. It has been hypothesised that attracting other Cliff Swallows benefits the signaller not because of anti-predator benefits, but because of the foraging benefits – increasing the number of individuals on the lookout for prey, which help the signaller keep track of a moving swarm of food.

Northern Ravens, which roost colonially in winter, also produce food calls and the hypothesised benefits are also related to foraging efficiency, though through slightly different means. Ravens are scavengers that often feed on large animal carcasses. Carcasses may be dispersed widely across the landscape, and can be hard to discover. Ravens produce food calls when they discover food, but not all individuals produce food calls in all contexts. Juvenile Ravens appear to produce food calls when they discover a carcass, and particularly when the carcass has already been claimed by a pair

↑

Cliff Swallow food calls may turn colonies into 'information centres' which increase the foraging success of all members of the colony.

↑

Northern Raven food calls may signal the age and sex of the caller, which could allow unfamiliar birds to make smart decisions about who to join for a meal and who to avoid.

←

Foraging House Sparrows produce 'chirrup' calls which attract other sparrows to a food source. They call less when they feel safer, such as when the food source is further away from the human collecting data on them.

of territorial adult Ravens. By attracting other foragers, the group of juveniles may be able to overcome the defences of the territorial adults. Even though the juveniles must share with each other, they now gain access to food that otherwise would have been unavailable.

The fact that food-related calls must have a benefit for the signaller can perhaps be made clearer by considering calling in Willow Tits. Willow Tits produce loud calls when an individual visits a new foraging patch, and the call attracts flock mates of both the same and different species to the callers' location. However, sometimes the Willow Tit immediately consumes the food it finds, and at other times it will cache, or hide, a food item, saving it for later. Willow Tits only produce food calls when they are eating the food they found – if they are caching food, they don't call. Obviously, they don't want to attract other birds to the location. Flock mates are important, but they are not more important than food.

7

COMMUNICATION IN A NOISY WORLD

All communication takes place in the context of noise, and birds need to make themselves heard above the noise of rival birds of the same species, other animals or the wind, for instance; just as a displaying male bird has evolved over time to make itself seen against the background of its habitat with colourful plumage and dramatic displays. However, in a rapidly urbanising world, birds face the additional challenges of transmitting their signals above the noise produced by humans. In this chapter, we look at how birds' songs, calls and plumage are being changed by human noise and pollution.

ENVIRONMENTAL NOISE

We began this book by discussing signal detection theory and how signals in birds evolve to transmit beneficial information, and we conclude by discussing how development of those signals is affected by noise.

Noise can be produced naturally in the environment and signals will evolve to stand out from that noise. However, anthropogenic environments present additional challenges to birds. Understanding how our activities influence communication is essential to understanding how human-altered landscapes are impacting birds.

For birds, anthropogenic noise can be extreme: sudden, unexpected and unpredictable in comparison with the background noises of the species and landscapes in the environment they've evolved in relation to over a long time. Any alteration of signals in response to noise generated by human activity needs to be immediate and may require changes in signal design and receiver sensory abilities that are not possible in the short term. Thus, anthropogenic noise can have significant impacts on signal detection and communication in birds. In addition, air pollution can also block or impede visual signals and act as noise in visual communication.

WHAT IS NOISE?

From the perspective of signal detection theory anything that is not signal is noise, and noise is a common part of the natural world – in fact, silence rarely exists. Naturally occurring noise can come from non-biological and biological sources. Environmental noise can include wind, rain, water, as well as communications by birds of the same species – for instance, the singing of other males during the dawn chorus – as well as sounds produced by different species, such as insects including cicadas.

The Red-rumped Parrot of Australia prefers open woodlands and is common in suburban parks and gardens.

How impactful the noise is on the ability of the signals to transmit information will depend on the signal-to-noise ratio (SNR). The higher the intensity of the signal relative to the noise, the more detectable the signal will be. As the SNR approaches 1, the signal becomes increasingly masked by the noise and at some point is no longer effective.

IMPACTS OF NATURAL NOISE ON VISUAL SIGNALS

Visual signals evolve to stand out in the environment. For instance, there is evidence to suggest that plumage colour used for signalling is best suited to the ambient light of the environment. In shaded forest, the dominant background colour is green. So the plumage

of forest birds used for visual signals in these environments will tend to contain more orange and red than birds found in open habitats to enhance the contrast with the background. Forest birds also have brighter, more reflective plumage to be more visible in low-light conditions.

In a study comparing plumage brightness and hue across numerous species pairs that lived in open habitats with those that lived in forested habitats, researchers reported a significant effect of the light environment on plumage signals that are consistent with optimising visibility in these lighting environments. Species that live in closed habitats, such as the Crimson Rosella, exhibited more colour in the long wavelength (orange and red) than closely related species, such as the Red-rumped Parrot, that live in open habitats. But they found that birds in open habitats were brighter or more reflective than forest birds, suggesting that birds in open habitats can take advantage of visual signals that travel long distances, with brighter and more reflective plumage being more visible at long distances.

The Crimson Rosella of southeastern Australia prefers mountain forests but is not an uncommon visitor to gardens. It has also been introduced to New Zealand and Norfolk Island.

IMPACTS OF NATURAL NOISE
ON ACOUSTIC SIGNALS

Acoustic signalling is impacted by attenuation (signal intensity decreases with distance) and degradation (signal quality is impacted by interference from the environment). Acoustic signals attenuate with distance, so that, at certain distances, they are no longer detectable from background noise. How quickly an acoustic signal attenuates depends on many environmental factors, such as atmosperic conditions, position of the singer relative to the ground (ground absorbs sounds) and habitat. How far a signal can travel depends on its acoustic properties. High-frequency sounds tend to be absorbed more readily by the ground or atmosphere – and so attenuate more quickly – than those at low frequencies. Higher-frequency sounds are also prone to more distortion from bouncing off surfaces and from turbulence in the atmosphere, including wind. Low-frequency sounds can travel especially long distances if they are produced near the ground, so the low-frequency (~100 Hz) booming sounds produced by a male Ruffed Grouse's wings from the forest floor, for example, can travel long distances.

Vocal sounds produced by birds are correlated with body size: the smaller the body size, the higher the frequency of vocalisations. Thus, most vocal sounds in birds are produced at frequencies higher than 100 Hz, usually in the range of 1000–10,000 Hz (1–10 kHz). Attenuation is a problem that might be mitigated by lowering the frequency of a signal, although due to the physical constraints on small birds (most songbirds are quite small), producing sounds at lower frequencies is not an option. Also, producing vocalisations that can travel long distances might not be helpful anyway. But there are ways that small birds can avoid degradation or distortion that are largely independent of frequency characteristics. Some patterns of song are impacted more by degradation such as reverberation, which occurs when sounds are reflected from solid surfaces to cause interference.

Forests are difficult acoustic environments, with surfaces including tree trunks and leaves reflecting and reverberating sounds in random directions and

degrading all frequencies of sound. The effect of reverberation is particularly pronounced in song patterns that include rapid repetition of elements, such as trills. Reverberation will cause interference in closely timed elements, making it difficult for receivers to make out temporal variation in song. Birds could potentially avoid degradation by avoiding patterns of song that are subjected to reverberation. Multiple studies that compared songs of species living in forested habitats with open habitats confirmed that the forest birds use songs that contain fewer trilled elements. It is possible that this is an evolutionary response in which the environment favours individuals that produce vocalisations which function best as signals and which, in songbirds, are easiest for juveniles to hear and learn.

Acoustic signals are best detected when they stand out from the background. For optimal detectability, background noise that masks or overlaps acoustic

The Three-wattled Bellbird of Central America deals with its noisy environment by making louder noise! It is one of four species of bellbirds that produce the loudest vocalisations of any birds, recorded at 125dB, which is as loud as a jet engine.

signals should be avoided. A common source of noise for singing birds is other birds singing. Birds wake up early to sing, the dawn chorus being a likely result of the best conditions for peak sound transmission occurring early in the morning, when the temperature is low and humidity is high. But this is also a time of peak noise because it is when most species sing. Birds will use various strategies to help their signal stand out from the noise.

One such strategy is to vary the timing of song so that there is minimal background noise. An extreme example of a temporal shift in singing activity can be found in Common Nightingales. When males are trying to attract mates, they will even sing during the night. Temporal adjustments to song can also occur more immediately. During playback experiments using the recordings of other species that naturally occur in the same areas as Nightingales, males avoided singing at the same time as the playback, presumably to avoid their song being masked by another singer.

The long-term effect of overlapping singing during the breeding season can be detected on an evolutionary timescale. A study of eighty-two species of birds that co-exist in the forests of the Amazon found that species which sang at the same time during the dawn chorus, in the same part of the forest, employed acoustic features in their songs that differed more than those sung by other species in different locations or at different times of the morning. Thus, birds can avoid some of the noise of other species by adjusting the timing of singing or the acoustic features of their song.

ATTENUATION AND DEGRADATION OF SONG AND THE EVOLUTIONARY CONSEQUENCES

Reverberation results in loss of signal integrity, especially in trills where elements are repeated. This figure shows the recordings made at different distances of two Carolina Wren songs broadcast from a loudspeaker in a room with many reflective surfaces. The original undegraded songs are shown in (a); (b) shows the same songs recorded at 4m, and (c) recorded at 9m. Reverberation has severely degraded the songs and this gets worse with distance. Blurring from one element to the next is characteristic of reverberation along with the loss of higher frequency notes. Reverberation is common in forested habitats which makes songs with rapidly repeated elements suboptimal signals here. Redrawn from Naguib (1996).

Carolina Wren

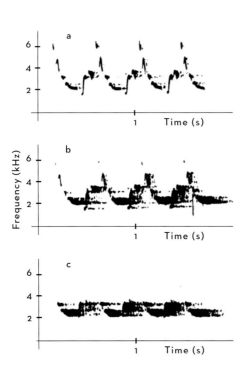

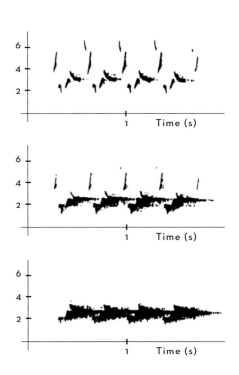

The evolutionary consequences of reverberation on song has been demonstrated in the Great Tit. This species is a habitat generalist and can be found living in forested and more open woodlands. In populations found in forests, males use songs with notes that are simple tones that do not exhibit frequency modulation. Their songs have evolved to minimise the effects of reverberation. The songs of birds from open woodlands include notes that are more complex. They include sounds at more and higher frequencies and contain elements that are repeated more rapidly. Redrawn from Krebs and Davies (1993).

Great Tit

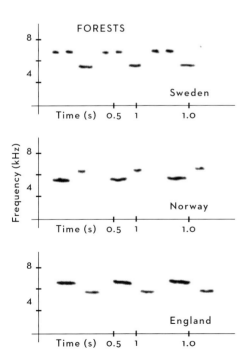

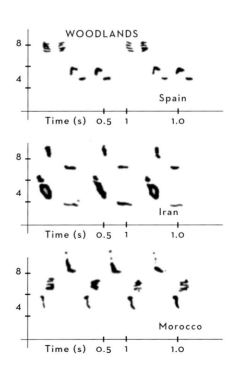

IMPACTS OF ANTHROPOGENIC NOISE ON VISUAL SIGNALS

Natural selection might favour plumage that blends in with the environment if camouflage is the goal. However, if a bird's plumage is meant to send a visual signal, then natural selection should favour it to stand out from the environment.

When researchers examined more than 1300 specimens of birds in the collection at the Field Museum in Chicago, ranging from the late 1800s to the present day, they found that there had been a significant impact from atmospheric pollution on plumage colour. Birds that were collected during times of industrialisation, before policy changes had resulted in the use of more efficient ways to burn coal, had plumage that appeared dark and dingy owing to sooty deposits in the feathers that were present in the atmosphere of the time. It is unknown how such discolouration might influence communication, but it is certainly possible that plumage discolouration resulting from air pollution has the potential to impact visual signalling.

In a 2010 study on how dirt influenced throat-breast plumage colour in European Starlings, researchers exposed male and female specimens to the dirt deposition from the air and compared them to specimens that were protected from dirt. After three weeks of exposure to air, there was a significant difference in reflectance across all wavelengths, but the effect was most significant on the ultraviolet- (UV-) reflecting plumage in males. The males impacted in this way could have their ability to signal reduced or need to invest more heavily in keeping soiled plumage clean.

In another study, carotenoid plumage in Great Tit nestlings was found to be influenced by the presence of a copper smelting facility in Finland. The nestlings that were reared closer to the copper smelter had less saturated carotenoid colour than the nestlings that were reared further away. In this case, the impact on plumage might be through the indirect effect of pollution on caterpillar larvae. Fewer carotenoid-rich caterpillar larvae were found near the copper smelting plant, which may have affected the availability of carotenoid to the nestlings.

The Common Starling of Eurasia has been introduced to North America, where it has thrived in highly urbanised areas.

Great Tits living in more urban areas may not have access to foods rich in carotenoids, which means their plumage is less yellow than that of their rural counterparts.

But it is also possible that ingestion of heavy metal pollution could influence carotenoid metabolism and deposition in growing feathers. Birds exposed to heavy metal pollution while they were being reared were less yellow and smaller, therefore the potential signalling function of their colourful plumage may remain intact. However, pollution which discoloured feathers could influence signalling by masking the brightness of plumage on an otherwise high-quality male. In that case, females choosing a mate based on bright plumage might miss pairing with a high-quality male whose plumage signal was masked by dirt. Pollution-masked plumage could have negative impacts on populations because high-quality males might not have the opportunity to mate, and females might not be receiving accurate information from plumage signals about male quality.

IMPACTS OF ANTHROPOGENIC NOISE ON ACOUSTIC SIGNALS

The levels of acoustic noise generated by human activity is astounding, and there is hardly a place left on earth that is not impacted – even the most remote locations may be affected by the noise from jet planes flying overhead. But birds that live near human activities are probably most impacted by the influence of anthropogenic noise on communication. Human noise can be loud and almost constant, as with the noise produced by vehicles travelling down a highway. It can also be loud and intermittent, as with noise from air conditioning and heating systems or from construction sites. Human noise can be constant but less loud, from people engaging in activities such as walking, running, talking or listening to music.

Human-generated noise typically has the highest amplitudes at low frequencies (1–2kHz) and can impact all types of acoustic communication in birds. Most research has focused on its effects on bird song. Some species such as European Robins will time their peak singing to avoid anthropogenic noise by singing more at night. But most birds do not make such temporal shifts and instead alter the frequency patterns of their songs. Researchers tend to find that as noise levels increase birds adjust song to use higher minimum frequencies.

It is hypothesised that birds exposed to anthropogenic noise are doing this to avoid being masked at the lower frequencies where urban noise is loudest.

However, there is some evidence to suggest that the degree of frequency shifting observed in urban birds may not make much difference to the detectability of their songs. So there might be more to the story of frequency shifting in urban singers. The results of a recent study on Common Blackbirds, a successful urban coloniser in Eurasia, found that males singing in the city used higher-frequency notes that were also louder than birds singing in forests. The researchers suggested that it is easier for the birds to sing louder at higher frequencies than at lower frequencies, enabling the urban singers to better avoid being masked by urban noise.

⬆
The ubiquitous European Robin does well in human-impacted environments throughout its range.

⬆
The Common Blackbird is well named as it is abundant across its natural range in Europe. It has also been introduced to Australia, where it is expanding its range.

Other studies have found a relationship between peak frequency and amplitude in bird vocalisations. In the laboratory, Budgerigars and Crested Tinamous produced vocalisations with both increased frequency and amplitude when they were exposed to noise. Tree Swallow nestlings have also been found to increase the frequency and amplitude of their begging calls in response to noise. The phenomenon of increasing pitch in response to noise is understood in humans as the Lombard Effect, but in the cases of the birds, the increases in frequency and pitch are an involuntary response to the increases in noise. In humans, the relationship between amplitude and frequency is attributed to the increased efficiency of energy transfer from the sound source to the air at higher frequencies than at lower frequencies. That is, a human can shout more loudly at higher frequencies than at lower frequencies. However, it remains unknown what mechanism explains the relationship between frequency and amplitude in songbirds, which have a more complicated sound production system. Nonetheless, in the lab and under natural conditions, there is a growing body of evidence to suggest that birds respond to noise by adjusting the frequency of their songs to be higher or using elements that contain higher minimum frequencies, and this may be a result of birds trying to sing more loudly to avoid their voices being hidden by the sounds of anthropogenic noise.

Tree Swallow nestlings produce loud begging calls from their nests, which can attract predators if parents do not hear them or attend to them quickly.

Elegant Crested Tinamous are native to South America and are related to the oldest extant lineage of flightless birds, the ratites, which includes ostriches and emus. However, tinamous are capable of weak flight.

White-crowned Sparrows are native to North America, where they can be found in both remote, rural locations such as the Sierra Nevada mountains and in human-impacted areas like Golden Gate park in San Francisco.

CONSEQUENCES OF ALTERED SIGNALLING

There are possibly expensive consequences for birds that adjust their signals to avoid or overcome noise. Male birds that alter frequency and amplitude of their songs may be increasing the energetic costs of singing. And one problem with adjusting songs to include higher-frequency notes is that the notes will be more prone to attenuation and degradation, especially in habitats that are complex, as is the case with urbanised areas.

Another problem is that producing sounds at higher frequencies may match less well with the acoustic sensitivities of the intended receivers, making it more difficult for other birds to hear the songs and detect information. For example, a male who adjusts song frequency may be paying an energetic cost that is not balanced by fitness benefits because females may not hear the signal, recognise the signaller as a bird of the same species or recognise the signaller as a high-quality male. Because frequency is related to body size, as we

have seen (see page 174), males that sing a higher-frequency song may be interpreted by competitors as smaller and less threatening – this may result in more confrontations that require the male's attention and energy. The same problems could apply to alarm and begging calls – they may not be properly heard or interpreted. So any changes that birds make to signalling in response to noise may be energetically costly and may not solve the problem of signal detection or even cause new problems.

While a lot of research has focused on the consequences for signallers, not much has been devoted to the consequences for receivers. What are the consequences of noise on receivers? And what are the consequences of the alterations of signallers in response to noise on receivers? One study on White-crowned Sparrows addressed both questions. Young White-crowned Sparrows trained with songs that included noise learned higher frequency songs than control birds. One noted consequence of song that was produced by males who learned songs in noise was that those songs were not only of a higher frequency but also that they exhibited lower vocal performance, a feature of song demonstrated to influence female mating preferences. Thus, receivers, in this case young males learning songs, are best able to learn songs that are less masked by anthropogenic noise. And, by learning songs that were easier to hear in noise they produced songs as adults that were not as effective in attracting females.

Another line of research on how males change song post position in response to noise suggested that receivers trying to hear signals might be paying increased costs to detect signals in noise. In a study on European Robins, researchers found that the anthropogenic noise level is positively correlated with song perch height. One interpretation of this observation is that, by singing from higher perches, males are trying to optimise their ability to hear their neighbours. However, by singing from higher perches, males might also be increasing the costs of signal detection by being more exposed to predation.

Other studies have shown behavioural responses to noise that may increase chances of detection. In a study on Great Tits, when a recording of noise was played at the nest box containing a nesting female, the female exhibited a delayed response to male song during the dawn chorus. Further, males responded by changing song perches to be closer to the nest box. Female reaction times returned to normal when males sang from song perches closer to the nest box.

Research so far suggests that anthropogenic noise is likely to have profound effects on receivers. Adjustments by senders to improve signal detection, such as changing the frequency of song, could also potentially impact a female's fitness if the adjustments affect her ability to judge male quality, causing her to choose a low-quality male or forgo mating altogether.

One could imagine that masking of signals can also cause significant problems for parents when listening for indications of hunger from begging nestlings. Increasing the frequency and amplitude of begging calls may impact a parent's ability to accurately determine the hunger or need of nestlings. Parents are likely to increase provisioning when begging calls are loud to reduce the probability of predation. Any increase in provisioning would be costly to adults.

More research is needed to understand how receivers are impacted by noisy environments, both in terms of changes made by signallers to increase their detectability and the receivers' ability to detect signals.

Altering signals, especially mating signals, to account for noise may have evolutionary or long-term consequences, including reproductive isolation and the formation of new, urban species. Birds are essentially colonising islands of urban habitat from nearby rural areas. If urban colonisers adapt to their new urban environments, and use mating signals that are not recognised by other populations, then the urban populations may have embarked on a unique evolutionary trajectory that could ultimately result in speciation. Species that are successful in urban colonies tend to exhibit parallel changes in signalling, such as frequency shifts, and in many other behaviours, too – they tend to become bolder and more aggressive, for example. It remains unknown whether changes in populations that live in urban areas are the result of natural selection or acclimatisation – more research is necessary to understand the complex impacts of urbanisation on birds.

IMPACTS ON CONSERVATION

Understanding the role of communication in the lives of birds is essential to conservation efforts. Conservationists who understand communication are better able to understand the needs of threatened and endangered species, and the impacts that certain activities may have on communication that are essential to species success. Studies on birds in North America and Europe have reported troubling findings that most species of birds are in decline. Decline in birds is attributed to anthropogenic habitat disturbance and destruction, pollution, global climate change, introduced species, building and window collisions, artificial light and noise. It is likely that changes in bird populations around the world are suffering similar declines. The rate of extinction in birds is well documented, with 181 species having gone extinct since 1500 (24 in the last 40 years) and 21 more species are categorised as critically endangered or possibly extinct. Birds likely are a canary in the coal mine (pun intended) for other less well-documented species like invertebrates and plants. But, maybe the enduring love for and fascination of birds will help generate awareness for all endangered species and bring about the political will necessary for long-term solutions that mitigate anthropogenic impacts on wildlife.

FURTHER READING

GENERAL BOOKS

Birds

Lovette, I. J., and J. W. Fitzpatrick, eds. 2016. *Handbook of Bird Biology*. New Jersey: John Wiley and Sons.

White, G. 1906. *The Natural History of Selborne*. London: Methuen.

Bird vocal communication

Catchpole, C. K., and P. J. Slater. 2008. *Bird Song: Biological Themes and Variations* (2nd edition). Cambridge, England: Cambridge University Press.

Kroodsma, D. E., and E. H. Miller, eds. 1996. *Ecology and Evolution of Acoustic Communication in Birds*. New York: Comstock Publishing.

Marler, P., and H. Slabbekoorn, eds. 2004. *Nature's Music: The Science of Birdsong*. San Diego: Academic Press.

Bird plumage

Hill, G. E., and K. J. McGraw, eds. 2006. *Bird Coloration: Mechanisms and Measurements. Vol. 1*. Cambridge, Massachusetts: Harvard University Press.

Hill, G. E., and K. J. McGraw, eds. 2006. *Bird Coloration: Function and Evolution. Vol. 2*. Cambridge, Massachusetts: Harvard University Press.

Animal communication

Bradbury, J. W., and S. L. Veherencamp. 2011. *Principles of Animal Communication* (2nd edition). Oxford: Sinauer Associates.

Searcy, W. A., and S. Nowicki. 2005. *The Evolution of Animal Communication: Reliability and Deception in Signaling Systems*. New Jersey: Princeton University Press.

Wiley, R. H. 2015. *Noise Matters: The Evolution of Communication*. Cambridge, Massachusetts: Harvard University Press.

Introduction: What is communication?

Dooling, R. J. 1992. Hearing in birds. *In* D. B. Webster, A. N. Popper and R. R. Fay, eds., *The Evolutionary Biology of Hearing*, pp. 545–559. New York: Springer.

Martin, G. R., K.-J. Wilson, J. M. Wild, S. Parsons, M. F. Kubke, and J. Corfield. 2007. Kiwi forego vision in the guidance of their nocturnal activities. *PLOS ONE* **2**(2): e198.

Chapter 1: Communication channels

Bonadonna, F., and A. Sanz-Aguilar. 2012. Kin recognition and inbreeding avoidance in wild birds: the first evidence for individual kin-related odour recognition. *Animal Behaviour* **84**(3): 509–513.

Bostwick Kimberly, S., M. L. Riccio, and J. M. Humphries. 2012. Massive, solidified bone in the wing of a volant courting bird. *Biology Letters* **8**(5): 760–763.

Douglas, H. D., A. S. Kitaysky, and E. V. Kitaiskaia. 2018. Odor is linked to adrenocortical function and male ornament size in a colonial seabird. *Behavioral Ecology* **29**(3): 736–744.

Dowsett-Lemaire, F. 1979. The imitative range of the song of the Marsh warbler *Acrocephalus palustris*, with special reference to imitations of African birds. *Ibis* **121**(4): 453–468.

Langmore, N. E. 1998. Functions of duet and solo songs of female birds. *Trends in Ecology & Evolution* **13**(4): 136–140.

Marler, P. 2004. Bird calls: their potential for behavioral neurobiology. *Annals of the New York Academy of Sciences* **1016**(1): 31–44.

Marler, P., and S. Peters. 1977. Selective vocal learning in a sparrow. *Science* **198**: 519–521.

Pepperberg, I. M. 1981. Functional vocalizations by an African grey parrot (*Psittacus erithacus*). *Zeitschrift für Tierpsychologie* **55**(2): 139–160.

Slater, P. J. B. 1989. Bird song learning: causes and consequences. *Ethology, Ecology & Evolution* **1**(1): 19–46.

Suthers, R., and S. A. Zollinger. 2004. Producing song: the vocal apparatus. *Annals of the New York Academy of Sciences* **1016**: 109–29.

Thorpe, W. H. 1954. The process of song-learning in the chaffinch as studied by means of the sound spectrograph. *Nature* **173**(4402): 465.

Westneat, M. W., J. H. Long Jr, W. Hoese, and S. Nowicki. 1993. Kinematics of birdsong: functional correlations of cranial movements and acoustic features in sparrows. *Journal of Experimental Biology* **182**: 147–171.

Wright, T. F. 1996. Regional dialects in the contact call of a parrot. *Proceedings: Biological Sciences* **263**(1372): 867–872.

Chapter 2: Male–female communication

Andersson, M. B. 1994. Sexual Selection. *In* J. R. Krebs and T. Clutton-Brock, eds., *Monographs in Behavior and Ecology*. New Jersey: Princeton University Press.

Andersson, S. 1989. Sexual selection and cues for female choice in leks of Jackson's widowbird *Euplectes jacksoni*. *Behavioral Ecology and Sociobiology* **25**(6): 403–410.

Bennett, A. T. D., I. C. Cuthill, J. C. Partridge, and E. J. Maier. 1996. Ultraviolet vision and mate choice in zebra finches. *Nature* **380**(4 April): 433–435.

Borgia, G. 1985. Bower quality, number of decorations and mating success of male satin bowerbirds (*Ptilonorhynchus violaceus*): an experimental analysis. *Animal Behaviour* **33**(1): 266–271.

Dalziell, A. H., R. A. Peters, A. Cockburn, A. D. Dorland, A. C. Maisey, and R. D. Magrath. 2013. Dance choreography is coordinated with song repertoire in a complex avian display. *Current Biology* **23**(12): 1132–1135.

Darwin, C. 1859. *On the Origin of Species*. 16th ed. Cambridge, MA: Harvard University Press.

Darwin, C. 1871. *The Descent of Man and Selection in Relation to Sex*. London: Murray.

Loyau, A., D. Gomez, B. Moureau, M. Théry, N. S. Hart, M. S. Jalme, A. T. D. Bennett, and G. Sorci. 2007. Iridescent structurally based coloration of eyespots correlates with mating success in the peacock. *Behavioral Ecology* **18**(6): 1123–1131.

McDonald, D. B., and W. K. Potts. 1994. Cooperative display and relatedness among males in a lek-mating bird. *Science* **266**(5187): 1030–1032.

McGlothlin, J. W., D. L. Duffy, J. L. Henry-Freeman, and E. D. Ketterson. 2007. Diet quality affects an attractive white plumage pattern in dark-eyed juncos (*Junco hyemalis*). *Behavioral Ecology and Sociobiology* **61**(9): 1391–1399.

Nowicki, S., and W. A. Searcy. 2004. Song function and the evolution of female preferences: why birds sing, why brains matter. *Annals of the New York Academy of Sciences* **1016**(1): 704–723.

Podos, J. 1997. A performance constraint on the evolution of trilled vocalizations in a songbird family (Passeriformes: Emberizidae). *Evolution* **51**(2): 537–551.

Searcy, W. A., and P. Marler. 1981. A test for responsiveness to song structure and programming in female sparrows. *Science* **213**: 926–928.

Chapter 3: Territoriality and dominance

Anderson, R. C., A. L. DuBois, D. K. Piech, W. A. Searcy, and S. Nowicki. 2013. Male response to an aggressive visual signal, the wing wave display, in swamp sparrows. *Behavioral Ecology and Sociobiology* **67**(4): 593–600.

Dey, C. J., J. Dale, and J. S. Quinn. 2014. Manipulating the appearance of a badge of status causes changes in true badge expression. *Proceedings of the Royal Society B: Biological Sciences* **281**(1775): 20132680.

Godard, R. 1991. Long-term memory of individual neighbours in a migratory songbird. *Nature* **350**(6315): 228–229.

Levin, R. N. 1996. Song behaviour and reproductive strategies in a duetting wren, *Thryothorus nigricapillus*: II. Playback experiments. *Animal Behaviour* **52**(6): 1107–1117.

Pryke, S. R., S. Andersson, M. J. Lawes, and S. E. Piper. 2002. Carotenoid status signaling in captive and wild red-collared widowbirds: independent effects of badge size and color. *Behavioral Ecology* **13**(5): 622–631.

Rico-Guevara, A., and M. Araya-Salas. 2015. Bills as daggers? A test for sexually dimorphic weapons in a lekking hummingbird. *Behavioral Ecology* **26**(1): 21–29.

Yasukawa, K. 1981. Song and territory defense in the red-winged blackbird. *The Auk* **98**(1): 185–187.

Chapter 4: Parent–offspring communication

Caro, S. M., A. S. Griffin, C. A. Hinde, and S. A. West. 2016. Unpredictable environments lead to the evolution of parental neglect in birds. *Nature Communications* **7**(1): 1–10.

Colombelli-Négrel, D., M. E. Hauber, J. Robertson, F. J. Sulloway, H. Hoi, M. Griggio, and S. Kleindorfer. 2012. Embryonic learning of vocal passwords in superb fairy-wrens reveals intruder cuckoo nestlings. *Current Biology* **22**(22): 2155–2160.

Jouventin, P., T. Aubin, and T. Lengagne. 1999. Finding a parent in a king penguin colony: the acoustic system of individual recognition. *Animal Behaviour* **57**(6): 1175–1183.

Krebs, E. A., and D. A. Putland. 2004. Chic chicks: the evolution of chick ornamentation in rails. *Behavioral Ecology* **15**(6): 946–951.

Chapter 5: Warning signals

Flower, T. 2011. Fork-tailed drongos use deceptive mimicked alarm calls to steal food. *Proceedings of the Royal Society B: Biological Sciences* **278**(1711): 1548–1555.

Méndez, C., and L. Sandoval. 2017. Dual function of chip calls depending on changing call rate related to risk level in territorial pairs of White-Eared Ground-Sparrows. *Ethology* **123**(3): 188–196.

Murray, T. G., J. Zeil, and R. D. Magrath. 2017. Sounds of modified flight feathers reliably signal danger in a pigeon. *Current Biology* **27**(22): 3520–3525.

Suzuki, T. N. 2014. Communication about predator type by a bird using discrete, graded and combinatorial variation in alarm calls. *Animal Behaviour* **87**: 59–65.

Templeton, C. N., E. Greene, and K. Davis. 2005. Allometry of alarm calls: black-capped chickadees encode information about predator size. *Science* **308**(5730): 1934–1937.

Chapter 6: Group life

Buhrman-Deever, S. C., E. A. Hobson, and A. D. Hobson. 2008. Individual recognition and selective response to contact calls in foraging brown-throated conures, *Aratinga pertinax*. *Animal Behaviour* **76**(5): 1715–1725.

McDonald, P. G., and J. Wright. 2011. Bell miner provisioning calls are more similar among relatives and are used by helpers at the nest to bias their effort towards kin. *Proceedings of the Royal Society B: Biological Sciences* **278**(1723): 3403–3411.

Price, J. J. 1999. Recognition of family-specific calls in stripe-backed wrens. *Animal Behaviour* **57**(2): 483–492.

Sewall, K. B. 2011. Early learning of discrete call variants in red crossbills: implications for reliable signaling. *Behavioral Ecology and Sociobiology* **65**(2): 157–166.

Suzuki, T. N., and N. Kutsukake. 2017. Foraging intention affects whether willow tits call to attract members of mixed-species flocks. *Royal Society Open Science* **4**(6): 170222.

Winger, B. M., B. C. Weeks, A. Farnsworth, A. W. Jones, M. Hennen, and D. E. Willard. 2019. Nocturnal flight-calling behaviour predicts vulnerability to artificial light in migratory birds. *Proceedings of the Royal Society B* **286**(1900): 20190364.

Chapter 7: Communication in a noisy world

DuBay, S. G., and C. C. Fuldner. 2017. Bird specimens track 135 years of atmospheric black carbon and environmental policy. *Proceedings of the National Academy of Sciences* **114**(43): 11321–11326.

Endler, J. A. 1993. The color of light in forests and its implications. *Ecological Monographs* **63**(1): 1–27.

Hunter, M. L., and J. R. Krebs. 1979. Geographical variation in the song of the Great tit (*Parus major*) in relation to ecological factors. *Journal of Animal Ecology* **48**(3): 759–785.

Moseley, D. L., G. E. Derryberry, J. N. Phillips, J. E. Danner, R. M. Danner, D. A. Luther, E. P. Derryberry. 2018. Acoustic adaptation to city noise through vocal learning by a songbird. *Proceedings of the Royal Society B: Biological Sciences* **285**(1888): 20181356.

Naguib, M. 1998. Perception of degradation in acoustic signals and its implications for ranging. *Behavioral Ecology and Sociobiology* **42**: 139–142.

Nemeth, E., N. Pieretti, S. A. Zollinger, N. Geberzahn, J. Partecke, A. C. Miranda, and H. Brumm. 2013. Bird song and anthropogenic noise: vocal constraints may explain why birds sing higher-frequency songs in cities. *Proceedings of the Royal Society B: Biological Sciences* **280**(1754): 20122798.

Wiley, R. H. 1991. Association of song properties with habitats for territorial oscine birds of eastern North America. *American Naturalist* **138**: 973–993.

INDEX

Note: page numbers in **bold** refer to information contained in captions.

PICTURE CREDITS

(t = top, c = centre, b = bottom, l = left, r = right)

Alamy Stock Photo: All Canada Photos 39b, 129, 142, 148, 155b, Arco Images GmbH 160, Auscape International Pty Ltd 67, Bill Brooks 168r, blickwinkel 28–29, 104, 107t, 112tl, Dave Watts 68, Genevieve Vallee 159, Gerry Matthews 82l, Images & Stories 107b, Ivan Kuzmin 33tr, Johner Images 50–51, Minden Pictures 116, National Geographic Image Collection 31t, 118, Nature Picture Library 124, 156, Raquel Mogado 154l, Rick & Nora Bowers 95l, Saverio Gatto 122, Zoltan Bagosi 183.

Getty Images: Ger Bosma 145, Javier Fernández Sánchez 97, Jonkman Photography 182, ©Juan Carlos Vindas 70–71, Paul Grace Photography Somersham 8, tracielouise 69, Vicki Jauron, Babylon and Beyond Photography 106, Zi Wang/EyeEm 109t.

Justin Schuetz: 117t.

Naturepl.com: Jen Guyton 12.

Shutterstock: A.von Dueren 112tr, Agami Photo Agency 9t, 40, 43t, 76, 88–89, 135tl, Agnieszka Bacal 26tl, 33c, 91t, Alan B. Schroeder 18, AlecTrusler2015 139t, Alekcey 85, Alex Babych 140–141, Alexandr Junek Imaging 172, Andre Valadao 117b, Andriy Blokhin 90–91, Angela N Perryman 143b, Anne Powell 173, Archaeopteryx Tours 113, Avvinte 86, Bachkova Natalia 2–3, bearacreative 55l, Belozersky 14b, Ben Schonewille 13t, Bildagentur Zoonar GmbH 57c, Bonnie Taylor Barry 39t, 53r, 84t, 94–95, Borislav Borisov 80t, Brian Lasenby 78t, Butler Stock Photography 101, Catzatsea 163b, Chris de Blank 156t, Christopher Becerra 174–175, Coatesy 110, COULANGES 23, Cristian Gusa 112b, Danita Delimont 87b, 166, David Kalosson 109b, David McManus 9b, 61t, David Tao 154r, Dee Carpenter Originals 60r, Delmas Lehman 151, Dennis W Donohue 99, EdwinWilke 135tr, Elina Litovkina 121t, Eric Isselee 36b, 57l, 177, Erni 81l, fernando sanchez 87t, FloridaStock 37, Fnach 27, FotoRequest 65l, Frank McClintock 15, Frode Jacobsen 20r, 96, Gino Santa Maria 152, goran_safarek 29t, Gualberto Becerra 90l, 164–165, guentermanaus 78c, Hannu Rama 52, Harold Stiver 26bl, 60l, 136r, Ian Dyball 25, Imogen Warren 91b, Jane Buttery 181, JasonYoder 31b, Jeff Caverly 138, JeremyRichards 123t, Jim Beers 64r, Jim Cumming 61b, Jiri Hrebicek 13b, 163t, Johan Swanepoel 77, Jukka Jantunen 144b, K Quinn Ferris 53l, Karel Bock 108, KarenGiblettPhotography 137, Keneva Photography 44–45, Kerry Hargrove 66, Kiril Kirkov 157, Laura Mountainspring 135b, Laurie E Wilson 20l, Leonie Ailsa Puckeridge 115r, Light Benders Visuals 33tl, Linda Burek 57r, Liz Miller 29b, LorraineHudgins 62–63, 126–127, Maciej Olszewski 36t, Martin Mecnarowski 11, 79, 162l, Martin Pelanek 64l, 81r, 93, Marty R Hall 167, Maslov Dmitry 56, Mauricio Acosta Rojas 48–49, Maximillian cabinet 119, Menno Schaefer 184, Michael Potter11 143t, Michsal Pesata 72, Michelle Sole 94t, Mircea Costina 80b, Mirko Graul 32, mlorenz 131c, Montipaiton 7, Moonborne 133, mvramesh 24, My Lit'l Eye 162r, Natalia Kuzmina 121b, 155t, Natalia Paklina 150t, Nellie Thorngate 114r, NeonLight 144t, Nick Pecker 35, 100, Nick Vorobey 169, 178, Ondrej Prosicky 147, 153, panda3800 59, Panu Ruangjan 46, Paul Reeves Photography 19, 82r, 95r, petrdd 33b, 58, piotreknik 180, R.C. Bennett 130, Ramona Edwards 38b, Ray Hennessy 54r, Reznichenko Sergey 105, Richard G Smith 98t, 131t, Roel Meijer 146, SanderMeertinsPhotography 55r, Sergey 402 123b, 150b, slowmotiongli 111, 179t, Soru Epotok 54l, Steve Byland 6, 38t, 65r, 83, Steve Dog Thompson 128, StockPhotoAstur 179b, StrippedPixel.com 170, Stubblefield Photography 78b, Super Prin 132, Syds.Pics 120, Takahashi Photography 98b, Tom Franks 84b, Tom Reichner 131b, Tony Campbell 156c, tryton2011 41, Ttshutter 136l, Vishnevskiy Vasily 30, 134, 168l, vladsilver 102, Wang LiQiang 114l, Wright Out There 115l, 139b, 161, Yakov Oskanov 43b, Yori Hirokawa 16.

Wikipedia Commons: Stripe-backed Wren/Cucarachero Chocorocoy (Campylorhynchus nuchalis brevipennis)/Fernando Flores/CC BY-SA 2.0, 158, juvenile male Indigo Bunting (Passerina cyanea) at Smith Oaks Sanctuary, High Island, Texas/Frank Schulenburg/CC BY-SA 4.0, 26r, Peruvian Warbling-antbird at Rio Branco, Acre, Brazil/Hector Bottai/CC BY-SA 4.0, 75, Carolina Wren (Thryothorus ludovicianus), Occoquan Bay National Wildlife Refuge, Woodbridge, Virginia/Judy Gallagher/CC BY-SA 2.0, 176.

All reasonable efforts have been made to trace copyright holders and to obtain their permission for the use of copyright material. The publisher apologises for any errors or omissions and will gratefully incorporate any corrections in future reprints if notified.